AF839794

Northwest Epidemiologic Enteric Disease Study

Awwa Research Foundation

About the Awwa Research Foundation

The Awwa Research Foundation (AwwaRF) is a member-supported, international, nonprofit organization that sponsors research to enable water utilities, public health agencies, and other professionals to provide safe and affordable drinking water to consumers.

The Foundation's mission is to advance the science of water to improve the quality of life. To achieve this mission, the Foundation sponsors studies on all aspects of drinking water, including supply and resources, treatment, monitoring and analysis, distribution, management, and health effects. Funding for research is provided primarily by subscription payments from approximately 1,000 utilities, consulting firms, and manufacturers in North America and abroad. Additional funding comes from collaborative partnerships with other national and international organizations, allowing for resources to be leveraged, expertise to be shared, and broad-based knowledge to be developed and disseminated. Government funding serves as a third source of research dollars.

From its headquarters in Denver, Colorado, the Foundation's staff directs and supports the efforts of more than 800 volunteers who serve on the board of trustees and various committees. These volunteers represent many facets of the water industry, and contribute their expertise to select and monitor research studies that benefit the entire drinking water community.

The results of research are disseminated through a number of channels, including reports, the Web site, conferences, and periodicals.

For subscribers, the Foundation serves as a cooperative program in which water suppliers unite to pool their resources. By applying Foundation research findings, these water suppliers can save substantial costs and stay on the leading edge of drinking water science and technology. Since its inception, AwwaRF has supplied the water community with more than $300 million in applied research.

More information about the Foundation and how to become a subscriber is available on the Web at **www.awwarf.org**.

Northwest Epidemiologic Enteric Disease Study

Prepared by:
Floyd J. Frost, Jr., **Twila R. Kunde**, and **Timothy L. Muller**
Lovelace Clinic Foundation
2309 Renard Pl., SE, Albuquerque, NM 87108

Sponsored by:
Awwa Research Foundation
6666 West Quincy Avenue, Denver, CO 80235-3098

Published by:

Awwa Research Foundation
Advancing the Science of Water®

IWA Publishing

DISCLAIMER

This study was funded by the Awwa Research Foundation (AwwaRF). AwwaRF assumes no responsibility for the content of the research study reported in this publication or for the opinions or statements of fact expressed in the report. The mention of trade names for commercial products does not represent or imply the approval or endorsement of AwwaRF. This report is presented solely for informational purposes.

Published by IWA Publishing, Alliance House, 12 Caxton Street, London SW1H 0QS, UK

Telephone: +44 (0) 20 7654 5500; Fax: +44 (0) 20 7654 5555; Email: publications@iwap.co.uk;
Web: **www.iwapublishing.com**

AwwaRF report number 91036F. Originally published by AwwaRF for its subscribers in 2004.
IWA Publishing version published 2005.

ISBN 1 84339 921 0

CONTENTS

TABLES

FOREWORD

The Awwa Research Foundation (AwwaRF) is a nonprofit corporation that is dedicated to the implementation of a research effort to help utilities respond to regulatory requirements and traditional high-priority concerns of the industry. The research agenda is developed through a process of consultation with subscribers and drinking water professionals. Under the umbrella of a Strategic Research Plan, the Research Advisory Council prioritizes the suggested projects based upon current and future needs, applicability, and past work; the recommendations are forwarded to the Board of Trustees for final selection. The foundation also sponsors research projects through the unsolicited proposal process; the Collaborative Research, Research Applications, and Tailored Collaboration programs; and various joint research efforts with organizations such as the U.S. Environmental Protection Agency, the U.S. Bureau of Reclamation, and the Association of California Water Agencies.

This publication is a result of one of these sponsored studies, and it is hoped that its findings will be applied in communities throughout the world. The following report serves not only as a means of communicating the results of the water industry's centralized research program but also as a tool to enlist the further support of the nonmember utilities and individuals.

Projects are managed closely from their inception to the final report by the foundation's staff and large cadre of volunteers who willingly contribute their time and expertise. The foundation serves a planning and management function and awards contracts to other institutions such as water utilities, universities, and engineering firms. The funding for this research effort comes primarily from the Subscription Program, through which water utilities subscribe to the research program and make an annual payment proportionate to the volume of water they deliver and consultants and manufactures subscribe based on their annual billings. The program offers a cost-effective and fair method for funding research in the public interest.

A broad spectrum of water supply issues is addressed by the foundation's research agenda: resources, treatment and operations, distribution and storage, water quality and analysis, toxicology, economics, and management. The ultimate purpose of the coordinated effort is to assist water suppliers to provide the highest possible quality of water economically and reliably. The true benefits are realized when the results are implemented at the utility level. The foundation's trustees are pleased to offer this publication as a contribution toward that end.

This study estimated the health benefits of filtering drinking water that was previously unfiltered but chlorinated. Disinfection now includes both ozonation as well as chlorination. The source water for this utility is derived from a well-protected watershed. The drinking water provided has not previously violated state or federal drinking water treatment standards and has not been involved in any prior waterborne disease outbreak. The new water treatment was expected to reduce the risks of waterborne disease. The benefits of filtration were measured by observing changes in waterborne disease risks associated with changes in water treatment. A secondary objective was to determine if waterborne disease risks are elevated in a community that uses unfiltered water from a well-protected surface water source.

Walter J. Bishop
Chair, Board of Trustees
Awwa Research Foundation

James F. Manwaring, P.E.
Executive Director
Awwa Research Foundation

ACKNOWLEDGMENTS

The authors of this report are indebted to the participating drinking water utilities and, most importantly, the many individuals who participated in this project. The assistance of the study participants was invaluable. Special thanks to our AwwaRF Senior Project Manager, Linda Reekie, and the Project Advisory Committee members: Deborah Levy, Ph.D., M.P.H., Centers for Disease Control and Prevention, Keith Osborn, United Utilities Water Ltd., and Stephen Hubbs, Louisville Water Company, for their assistance and guidelines on this project. The authors would also like to thank Rebecca L. Calderon, Ph.D. and Christina Peterson, Ph.D. of the United States Environmental Protection Agency, along with Gunther F. Craun, P.E., M.S., M.P.H. and Michael F. Craun, P.E., M.S. of Gunther F. Craun & Associates for their collaborative efforts with the study design and engineering evaluation and Melissa Roberts, M.S. of Lovelace Respiratory Research Institute for her expertise in SAS programming and statistical analyses

We would also like to thank the field staff for their assistance and efforts in this project: Lucy Harter, Debby Nagusky, Jane Lee, Regi Lyne-Carter, Amy Benegston, Carla Camou, Mickey Hardin, Karen Seitz, and Cindy Green.

EXECUTIVE SUMMARY

RESEARCH OBJECTIVES

The primary objective of the study was to estimate the level of infectious gastrointestinal disease attributable to drinking water. The study enrolled three groups of individuals served by different water sources. The approach was to compare rates of enteric illnesses in the three cohorts before changes were made in drinking water treatment at one site with enteric illness rates after the change occurred. A secondary objective was to determine if drinking-water-attributable endemic levels of gastrointestinal diseases are elevated in a community that uses unfiltered water from a well-protected surface water source. The approach for this objective was to compare illness rates between the surface water sites and the ground water site before the intervention.

APPROACH

This is an epidemiological study of two populations using unfiltered but chlorinated surface-derived drinking water and one population using ground water. In the fall of 2000, filtration and ozonation were initiated at one unfiltered surface water source. No changes were made in the sources or water treatment at the other locations. Cohorts of individuals were assembled from each population and illnesses were tracked using daily illness diaries in three cohorts for six months prior to initiation of filtration (Phase 1) and the same six-month time frame the following year after the implementation of filtration (Phase 2). The three sites included: Site A that was affected by the water treatment change in the fall of 2000, Site B that was served by chlorinated, unfiltered surface water for the entire study period, and Site C that was served by chlorinated and uncontaminated ground water. All sites before and after the treatment changes met the Safe Drinking Water Standards. Sera collected from adults during each phase of the study were analyzed for antibody responses to *Cryptosporidia* and *Giardia* by Western blot. Stools collected in Site A during both phases of the study were tested using a polymerase chain reaction (PCR) test for Microsporidia, *Cryptosporidium,* and *Giardia,* along with *Cryptosporidium* IgA antibody by ELISA. Daily illness diaries collected before and after initiation of filtration of the unfiltered drinking water supply were used to estimate the risk of waterborne gastroenteritis. Other information was also collected from Sites A and B to support the study efforts. This information collected during both phases included: 1) emergency department visits, 2) school absences, and 3) water quality data.

CONCLUSIONS

This study had two objectives. The first was to estimate drinking-water-attributable endemic levels of gastrointestinal diseases. No reduction in illness rates was observed for the intervention site and we found no changes in serological responses suggestive of a reduced occurrence of *Cryptosporidium* infection. A secondary objective was to determine if drinking-water-attributable endemic levels of gastrointestinal diseases are elevated in the communities using unfiltered surface-derived drinking water from well-protected sources. We did not observe higher levels of illness rates or serological responses in users of unfiltered but chlorinated surface water versus ground water users. This may have been due to the high quality of the surface source water. The two watersheds supplying surface drinking water were well protected, receiving no human or domestic animal fecal waste.

The study found lower rates of reported diarrhea, other gastrointestinal illness, and other illness in Phase 2 than Phase 1 for all study sites ($p<0.01$). However, there was no greater decline in the occurrence of diarrheal or other gastrointestinal illnesses in the intervention site than the two control sites. In fact, compared to Site B, Site A (the intervention site) and Site C (groundwater control site) experienced a relative increase in diarrhea and highly credible gastrointestinal illnesses (as defined by Payment et al. 1991a, 1991b) in Phase 2 ($p<0.05$). For Site A, the rates of diarrheal and highly credible gastrointestinal illness events lasting 2 or more days increased in Phase 2 ($p<0.01$). Changes from Phase 1 to Phase 2 in the occurrence of vomiting, nausea, or abdominal cramping (i.e. gastrointestinal illness), and all other illnesses were unrelated to the site. Reasons for an increase rather than a decrease in diarrhea illness are unclear. In the stool samples, the study found intestinal IgA responses to *Cryptosporidium* in a number of individuals, however, PCR analysis failed to detect additional cases of infection. The rate of these detections was unrelated to study site or phase.

CHAPTER 1
INTRODUCTION AND BACKGROUND

INTRODUCTION

Despite major improvements in drinking water treatment and an aggressive regulatory approach, outbreaks of waterborne illness continue to occur in the United States (Craun et al. 1996a, Craun and Calderon 1997a). Since 1990, several large *Cryptosporidium outbreaks* have occurred, including a 1993 Milwaukee outbreak that caused an estimated 403,000 cases of illness, 1,000 hospitalizations, and 100 deaths (MacKenzie et al. 1994, Blair 1994, Hoxie et al. 1996). Other waterborne outbreaks from *Shigella* and enterohemorrhagic *E. coli* contamination of the water supply have also caused deaths and severe illness (Craun 1992). Despite efforts to better identify and investigate waterborne outbreaks, outbreaks and illnesses remain under reported and a significant number of the reported outbreaks result in illnesses of undetermined etiology, with symptoms such as nausea, vomiting, diarrhea, and fever (Craun and Calderon 1997a).

In addition to unreported outbreaks, there is increasing evidence of unrecognized endemic waterborne illness, and this may be a greater problem than outbreaks (Frost et al. 1996). Unfortunately, little epidemiological data are available to estimate the magnitude of these risks. Morris and Levin (1995) estimated that 1.8 million cases of waterborne disease and 1,800 deaths occur annually. These estimates are very uncertain because they were not based on epidemiological studies. Few such studies have been conducted. Randomized prospective epidemiological studies in a community where no outbreaks were reported and drinking water met or exceeded current standards found that 14% to 35% of gastrointestinal illnesses could have been waterborne (Payment et al.1991a, Payment et al.1991b). A follow-up study suggested that distribution system contamination might have been partly responsible for endemic gastroenteritis (Payment et al. 1997). Other studies suggest that 40% to 70% of endemic giardiasis may be associated with unfiltered surface water supplies (Chute et al. 1987, Birkhead and Vogt 1989, Fraser and Cook 1991). However, recent well-designed and rigorously conducted studies (Hellard et al. 2001, Colford et al. 2003) have not been able to detect evidence of endemic waterborne disease risks.

BACKGROUND

Carefully designed epidemiological studies are needed to accurately estimate the risks of endemic waterborne infections and disease (Frost et al. 1996). It should also be recognized that many waterborne infections result in mild illness or may be asymptomatic and that inhalation can also be a route of exposure for several waterborne pathogens. Serological techniques offer the promise of improving risk estimates for waterborne infection by protozoa and several viruses in the absence of clinically recognized disease (Frost et al. 1997). A recent serological study funded by the Awwa Research Foundation (AwwaRF) found an elevated risk of *Cryptosporidia* infection for consumers of filtered and chlorinated surface water compared to consumers of ground water but found no evidence of an outbreak or excess reported illnesses, no violations of current drinking water quality standards, and no evidence of excess illness associated with the serologically identified *Cryptosporidium* infections (Frost et al. 1998c). Serological studies and evaluations of waterborne cryptosporidiosis outbreaks (Craun 1996b) suggest that chronic waterborne *Cryptosporidium* exposures and infections may reduce the occurrence or severity of illness (Frost et al. 1997). Studies in the United Kingdom (Meinhardt et al. 1996) and laboratory experiments using human volunteers (Moss et al. 1998, Okhuysen et al. 1998a) support these findings.

A number of methodological issues have been raised in investigating endemic waterborne disease, including study design. An EPA and CDC workshop discussed the types of epidemiological studies that should be considered to estimate the attributable-risk of waterborne disease (Hunter et al. 2001). CDC preferred an experimental study design in which people are enrolled in a blinded trial and receive water with (experimental) or without (control) additional treatment at the tap. Although this design is successful for clinical trials, tap water exposures are pervasive, making it difficult to ensure that an experimental group will be completely unexposed to tap water. Approximately 20% of tap water is consumed outside the home (Payment et al. 1991a), which may significantly reduce the power of home-based clinical trials to detect waterborne disease risks. Although 18% and 31% of two New York City cohorts denied any tap water consumption, 96% and 98% responded positively when questioned about alternate tap water ingestion (e.g., restaurants, friend's homes, etc.) (Davis et al. 1998). Incorrect exposure assignment can result from visiting other houses, travel, or use of bottled water. This is a problem with most study designs, including the approach used in this study, but it is more of a problem if only a small fraction of the water consumed in a community is changed, as would occur in the randomized household intervention studies.

The Community Intervention Design has the advantage that all of the drinking water for a community changes at one time, usually when a new water source is obtained or a new water treatment system is installed. All residents of an area receive the new treatment of the drinking water at the time of the intervention. This design eliminates both direct pathogen transmission from drinking water as well as secondary spread of the waterborne pathogen.

A problem with both the randomized household intervention and the community intervention designs is that neither considers the effect of protective immunity from prior exposure to the drinking water. After a period of consuming water from a particular source, it is likely that the individual develops an immune response to the organisms that may cause illness. If so, improved water treatment may not provide reduced illness risk to individuals who have already developed a protective response to the organisms. One would falsely conclude from studies that the improvement in treatment was unrelated to reduced illness risks simply because the study examined people no longer susceptible to disease from the waterborne organisms.

For all of the study designs, questions have also been raised about which health outcomes are of interest. Outbreak investigations are based on occurrences of illness clusters epidemiologically associated with drinking water. The development of serological tools to identify infections from parasites (*Cryptosporidia* and *Giardia*) and viruses may increase the sensitivity of epidemiological studies of waterborne disease by more completely identifying infections. Changes in serological response to pathogen antigens suggest that infection has occurred. If a large fraction of waterborne infections do not result in clinically detectable illness (e.g., mild diarrhea), then the illnesses that are detected will not accurately reflect the magnitude of waterborne risk and can misrepresent important risk factors for infections (Frost et al. 1997). Even though serological studies have proven to be effective for assessing waterborne *Cryptosporidium* infection risk, questions remain about the appropriateness of some serological markers for epidemiological studies (Okhuysen et al. 1998a, 1998b). For example, if infection were defined as the occurrence of a pathogen in the stool or in body tissues, the absence of either illness associated with seroconversion and the absence of pathogens in the stool or other clinical specimens would suggest these markers do not reflect an increased occurrence of infection (Okhuysen et al. 1998a, 1998b). This argument raises questions about the definition of infection. Is evidence of seroconversion in the absence of either illness or the identification of a pathogen sufficient to conclude that infection has occurred? Although mild illnesses may well be of little health significance, for some diseases, such as polio, assessing serological responses in the absence of illness was critical to understanding the patterns of disease occurrence. Serological studies offer several advantages in helping to assess waterborne infection risk for persons exposed to certain waterborne pathogens. The most important is that large populations of ill persons need not be studied. A variety of published and unpublished studies indicate that people respond serologically to *Giardia* and *Cryptosporidia* infection, but additional studies are needed to better interpret the significance of serological data for assessing waterborne risks.

A paired city study of *Cryptosporidia* infection was previously conducted based on changes in serological responses to antigens in one city that uses filtered surface water and another city that uses a high quality ground water source (Frost 1998). Rates of seroconversion were compared and higher rates were observed for the surface water city. This paired city design has been repeated using a second pair of cities with the same findings (Frost et al.#1 in press.). Both studies examined chlorinated and filtered surface-derived drinking water where the source has considerable human and animal sewage contamination. In addition, the authors conducted a pilot study in the New England area that examined serological responses to *Cryptosporidia* antigens in college students several months prior to installing drinking water filtration and several months afterwards. Very low levels of serological responses were found in both surveys and there was no evidence of reduced serological responses among these college students after the initiation of drinking water filtration. Preliminary results from another of these studies based on the National Health and Nutritional Examination Study (NHANES) sera also suggest that cities using source water that has little human or animal sources of fecal contamination have reduced levels of serological response to *Cryptosporidia* antigens (Frost et al. #2 in press). These studies suggest that source water quality may be an important predictor of the risk of endemic waterborne infection. However, the relationship between endemic infections and the risk of illness is unclear.

RESEARCH OBJECTIVES

The primary objectives of this study were to estimate endemic levels of diarrhea and gastrointestinal diseases in a population served by either an unfiltered surface-derived drinking water obtained from a well-protected watershed or a groundwater source and to determine if these illness rates declined following initiation of filtration and ozonation at one of two surface water sites. Rates of illnesses before and after implementation of filtration were estimated for the site where the intervention occurred (Site A) and for the other two sites. The second site (Site B) was another area of the same city served by another unfiltered but chlorinated drinking water source. This site served as a surface water control. The third site (Site C) was distant from the other two sites and was served by uncontaminated ground water. This site served as a groundwater control. The controls were used to determine if there was a difference between the surface water cohort and the groundwater cohort during Phase 1 and if there was a difference between the two surface water cohorts and ground water cohort during Phase 2.

Cohorts of individuals were enrolled from each site and illness diaries were maintained in a period before and a similar period after filtration and ozonation was initiated at the intervention surface water site.

This study compared the occurrence of infectious enteric diseases and infections from selected pathogens between populations in Sites A, B, and C. The surface water sources for Sites A and B had no human and few animal sources of fecal contamination. In November 2000, the treatment process for Site A changed from chlorinated and unfiltered to ozonated, filtered, and chlorinated.

A primary goal of the study was to estimate endemic levels of diarrhea and gastrointestinal diseases. This was done by measuring changes in diarrheal and gastrointestinal disease risks associated with changes in water treatment. A secondary objective was to determine if waterborne risks are elevated in a community that uses unfiltered water from a well-protected surface water source. This was done by comparing illness levels in users of surface and ground water before the intervention. The risk of gastroenteritis, *Giardia, Cryptosporidia,* and enteric viral infections attributable to drinking water in each of the cohorts were estimated in the 6-month period before and in a comparable 6-month period after the changes in the water disinfection practices and addition of filtration. This allowed an assessment of the benefit accrued from changes in drinking water treatment. School absenteeism, water quality and emergency department visits in the study areas were tracked during the same time frames.

CHAPTER 2
METHODS

SUBJECTS

Eligibility

This study recruited families receiving their drinking water from one of three sources [either one of two surface water sources or a ground water source] and had either a child between the ages of 2 through 10 or an adult who was 65 years or older. Family members must have consumed municipal drinking water, lived in the area for at least 6 months, and been of good physical and mental health. Residential areas were selected for recruitment potential through identification of the source of their drinking water. Recruitment was done through mailings targeted at zip codes that were served by one of the two surface water sources (Sites A and B) or water districts (Site C) from June through September 2000.

Table 2.1
Phase 1 recruitment by site and cohort

		Families Enrolled			% Recruitment
Cohort	Fliers Mailed	With Seniors	With Children	Total	(#Families/#Fliers)
Site A	72,533	132	225	357	0.5%
Site B	17,309	97	115	212	1.2%
Total Sites A & B	89,842	229	340	569	0.6%
Site C	18,403	83	104	187	1.02%
Total	10,245	312	444	756	0.7%

As shown in Table 2.1, there was a higher overall recruitment rate per mailed flier in Site C as compared to the average of Sites A and B. This could have been due to the use of address labels for utility customers provided by the water utilities in Site C. These labels contained the names of the consumer, which made the flier more personable. In contrast, the labels used in Sites A and B were addressed to "Resident". However, when results from Site A were separated from those of Site B, the higher recruitment rate for Site B negates this 'impersonal' theory. The recruitment rates were comparable to those of a CDC intervention study. (Calderon, personal communication, 2002)

Enrollment

Our enrollment target for Site A was 275 families with children and 137 with seniors. In Site B, our enrollment target was 167 families with children and 83 with seniors. In Site C, our target was 167 families with children and 83 families with seniors. These numbers were based on the need to detect a 15% reduction in the occurrence of enteric illness in Site A during Phase 2. The sample size calculation was based on an expected occurrence of highly credible gastrointestinal events in a prior study (Payment et al. 1997) for ages 0 to 5 years, 6 to 20 years, 21 to 49 years and 50 years and older. With reduced reporting of illness in Phase 2, the actual power of the study would be lower than estimated.

The identified population received recruitment fliers through the mail. If they were interested in participating in the study, they were asked to either call the field offices to complete the screening process or to complete the postcards as "Interested" and mail them to the field office. Subjects were then screened over the telephone using the *Screening Questionnaire* to determine eligibility. If the family met the enrollment criteria described earlier, an appointment was made for a study enroller to visit the family in their home to describe the project in more detail and verify that the family was still interested in participating in the project. If the family agreed, *Consent forms* were collected for all of the participating adults aged 18 years and older and *Assent forms* were collected for all children between the ages of 7 and 17. The main contact for the family was identified. This person was responsible for completing the forms for the family and would be the contact if there were any questions about the submitted forms. This person was designated as the "main adult". During enrollment, an *Enrollment Questionnaire Part A* form was completed by the main adult, detailing miscellaneous demographic information for that person. *Enrollment Questionnaire Part B* was completed for each family member. This form detailed demographic information for each family member, including occupation, general water consumption, medications, etc. All of the questionnaires had been used in previous EPA studies.

Participating families were assigned a four-digit family identification number with each participating family member being assigned a two-digit individual identification number. The family identification number was linked to the study site (1000-1999 for Site A, 2000-2999 for Site B and 3000+ for Site C). For the families with children, the "01" individual identification number was given to the main person of the family and the "02" was given to the first qualifying child. For the families with seniors, "01" was given to the main contact who was usually the main person. The families were categorized as "Families with Children" or "Families with Seniors" and by their location: Site A, Site B and Site C. Site A was the intervention site with the change in water treatment in Phase 2 of the surface water. Site B was the surface water source with consistent water treatment, and Site C was the ground water source with consistent water treatment.

Actual enrollment in Site A was 225 families with children (81.8% of the target) and 132 families with seniors (95.6% of the target). In Site B, actual enrollment was 115 families with children (68.9% of the target) and 97 families with seniors (116.9% of the target). In Site C, actual enrollment was 104 families with children (62.3% of the target) and 83 families with seniors (100% of the target).

Phase 2 Re-enrollment

A total of 172 families disenrolled from Phase 1 to Phase 2. The distribution of enrollment in Phases 1 and 2 is given in Table 2.2.

Table 2.2
Phase 2 re-enrollment

Site	Enrolled in Phase1I	Enrolled in Phase 2	% Re-enrolled from Phase 1
Families with Children			
Site A	225	170	75.5%
Site B	115	89	77.4%
Site C	104	71	68.3%
Families with Seniors			
Site A	132	109	82.5%
Site B	97	84	86.6%
Site C	83	61	73.5%
Total	756	584	74.3%

The main reason given by families with children for dropping out of the study was the families were busy and they had inadequate time to complete the diaries. Families with seniors also expressed concerns about the time required to complete the forms, but also mentioned travel plans and new health problems. Death also caused some disenrollment. The majority of the people who disenrolled were non-compliant with data collection. It is, therefore, difficult to determine whether the people who disenrolled from the study would have been comparable to those who remained.

Health Tracking

The main person from each family was responsible for completing the forms for the entire family. The Daily Diary (DD) was used to record the general health of each family member. This form recorded if the person was either out of town, well, or had diarrhea [loose/watery stools], vomiting, nausea, or other symptoms of colds, etc. An Illness Event Questionnaire (IEQ) was to be completed if the person had any of the following symptoms: 2 or more episodes of diarrhea, vomiting, or nausea or abdominal cramping. This form also asked whether the subject had traveled, eaten out, visited various amusement sites, had contact with animals, had contact with infants, or had contact with sick individuals, along with other potential risk factors. In addition, this form asked for the subject's opinion as to what made them ill. These forms were mailed in weekly, with follow-up from field staff if there were any questions concerning the content or lack of an IEQ.

The Water Consumption Questionnaire (WCQ) asked for the amount of water consumed for each family member in one day. This form was completed twice each phase [early and late in each Phase], at approximately the same time each year.

All of the questionnaires used in this study had been approved by the U.S. Office of Management and Budget and used in prior EPA field studies. All data collection instruments were approved by multiple human ethics or institutional review boards prior to use.

Statistical Analysis of the Daily Diary Data.

In epidemiological research, many exposures and outcomes have a complex hierarchical structure. This is especially the case for data collected from multiple members of the same family. Members of the same family often share both exposures and other disease risk factors and may tend to be more similar than people from different families. Most statistical analyses assume the data are 'independent, and identically distributed'. Ignoring the hierarchical or clustering characteristics of data can lead to either wrong or misleading conclusions. In this study, there are two categories in which the outcome may be influenced:

within-cluster effects: within the family

hierarchical effects: grouping of families into areas

SAS statistical software version 8.2 was used to aggregate data within families and families within water treatment service areas. For hierarchical modeling, MLwiN statistical software, version 2.0 (Centre for Multi-level Modeling, Institute of Education, London, England UK) was used. In modeling uncommon occurrences, such as diarrhea, gastrointestinal and other events where low values of expected occurrences are a source of more variation in the calculated ratios, counts of occurrences were modeled using a Poisson response. To better understand the nature of the disease, lengths of illness events were identified as those lasting either one day or more than one day. The majority of events reported lasted only one day.

Compensation

Families with children were paid \$20 per month of completed diaries. Families with senior citizens were paid \$15 per month of completed diaries. Families were also paid \$15 for each blood or stool sample that was submitted. Additionally, families received an EPA beanie toy upon completion of certain stages of the project, for a total of 3 toys per family.

In between Phase 1 and 2, newsletters, generated jointly by EPA and Lovelace Clinic Foundation (LCF), were sent to the families to encourage continued participation in the study. These newsletters contained general environmental and study information.

BIOLOGICAL SAMPLES

At the time of enrollment, healthy adult subjects were asked if they would contribute either a stool and/or a blood sampling during each year of the study. Children were asked if they would contribute a stool sample. These samples were collected when the subject was healthy, not when the subject was ill with gastroenteritis. The subjects also completed a Healthy Subject Blood/Stool Questionnaire (HSQ). This form described contact with animals, travel, people, etc. in the past month. An illness event biological sampling program was deemed too cumbersome to complete at the time due to time constraints with the sample collection.

Blood Samples

Blood samples were collected from adults in all three cohorts. The subjects were instructed to go to a participating facility to have their blood drawn. The sample was processed into two sera aliquots. These aliquots were shipped separately to Lovelace Clinic Foundation in Albuquerque, NM. They were tested for *Cryptosporidia* and *Giardia* IgG by Western blot.

Table 2.3 details the number of blood samples received during each Phase. One hundred and eighteen subjects from Site A and B gave blood samples in both Phases. Sixty-six subjects from Site C gave blood samples in both Phases.

Table 2.3
Blood sample donation [adults only]

	Phase 1	Phase 2
Site A-Children	73	73
Site A-Seniors	59	54
Site B-Children	20	13
Site B-Seniors	52	26
Site C-Children	28	36
Site C-Seniors	55	60

Cryptosporidium Western Blot Analysis

As mentioned earlier, one of the problems with most of the designs is that the effect of protective immunity is not considered. A protective immune response usually develops for pathogens such as *Cryptosporidium* and *Giardia* following repeated infections. If so, most individuals in the community are either less susceptible or no longer susceptible to illness from infection. However, serological responses will occur following repeated infections, even in the absence of illness. The occurrence of a waterborne infection can be as important as the occurrence of a waterborne illness.

A variety of approaches to detecting these serological responses have been proposed. Usually, tests with more specific responses to the pathogen of interest are considered the best tests. Also, specific responses can be achieved in a variety of ways. Issues of cost, reproducibility and reliability of the test are all important considerations. For this study, sera were analyzed by immunoblot to measure serological response to the 15/17- and 27-kDa antigen groups. These responses are considered to be specific to *Cryptosporidium* and the methods have been described in a previous study (Frost et al. 1998a). Both protein groups rise rapidly after exposure to the oocysts. The 15/17-kDa marker levels appear to return to normal more quickly than the 27-kDa marker.

Cryptosporidium protein is separated by Sodium Dodecyl Sulfate Polyacrylamide gel electrophoresis (SDS-PAGE) using a 15% separating gel in a continuous buffer system. The protein is electrophoretically transferred out of the gel and onto a polyvinylidene fluoride (PVDF) membrane. The PVDF membrane is blocked with PBS/0.3% Polyoxyethylenesorbitan Monolaurate (Tween 20) for 30 minutes and placed in an immunoblot screening device. The device creates leak-proof channels to aid in the analysis of different sera samples for the presence of human anti-*Cryptosporidium* antibodies. Each isolated chamber is exposed to human sera at a 1:50 dilution in PBS/0.3% Tween 20 overnight at 4° C. The bound primary human anti-*Cryptosporidium* antibodies are reacted with IgG1, IgG2, and IgG4 biotinylated mouse anti-human antibodies at a 1:500 dilution in PBS/0.3% Tween 20 for 60 minutes. Bound secondary antibodies are incubated with streptavidin alkaline phosphatase at a 1:1000 dilution in PBS/0.3% Tween 20 for 60 minutes. The bound antibodies are visualized with 5-Bromo-4-

Chloro-3-Indolyl Phosphate as substrate and Nitro Blue Tetrazolium as the chromogen. The intensity of the serological responses to the 15/17- and 27-kDa antigen groups on the immunoblots are digitally analyzed by an IS-2000 Digital Imaging System (Alpha Innotech). The image is captured using a high-performance Charge Coupled Device (CCD) camera and the system calculates the pixel density of the manually selected band of the immunoblot. This allows the intensity of the serological response on the immunoblot to be quantified.

The results for each specimen were standardized by taking the ratio of the response intensity for the unknown sample to the response intensity for a positive control serum contained on each blot. The IgG positive control serum was derived by mixing sera from several individuals who had strong serological response to both antigens. Results were reported as the ratio of the unknown to the positive control for the specific blot.

The serological tests and the quality control procedures used were described in detail in a prior AwwaRF publication on a paired city study (Frost 1998). The quality control procedures for the *Giardia* Western blot analysis were similar.

Giardia Western Blot Analysis

Sera collected from adults were analyzed by immunoblot to measure IgG serological response to the 31-kDa-antigen group as described by Taylor et al. (1987). *Giardia* protein is separated by Sodium Dodecyl Sulfate Polyacrylamide gel electrophoresis (SDS-PAGE) using a 15% separating gel in a continuous buffer system. The protein is electrophoretically transferred out of the gel and onto a polyvinylidene fluoride (PVDF) membrane. The PVDF membrane is blocked with PBS/0.3% Polyoxyethylenesorbitan Monolaurate (Tween 20) for 30 minutes and placed in an immunoblot screening device. The device creates leak-proof channels to aid in the analysis of different sera samples for the presence of human anti-*Giardia* antibodies. Each isolated chamber is exposed to a 1:50 dilution of human sera in PBS/0.3% Tween 20 overnight at 4°C. The bound primary human anti-*Giardia* antibodies are reacted with IgG1, IgG2, and IgG4 biotinylated mouse anti-human antibodies at a 1:500 dilution in PBS/0.3% Tween 20 for 60 minutes. Bound secondary antibodies are incubated with streptavidin alkaline phosphatase at a 1:1000 dilution in PBS/0.3% Tween 20 for 60 minutes. The bound antibodies are visualized with 5-Bromo-4-Chloro-3-Indolyl Phosphate as substrate and Nitro Blue Tetrazolium as the chromogen. The intensity of the serological responses to the 31-kDa antigen groups on the immunoblots were digitally analyzed by an IS-2000 Digital Imaging System (Alpha Innotech). The image is captured using a high-performance Charge Coupled Device (CCD) camera and the system calculates the pixel density of the manually selected band of the immunoblot. This allows the intensity of the serological response on the immunoblot to be quantified.

As for the *Cryptosporidium* Western blot, the results for each specimen were standardized by taking the ratio of the response intensity of the unknown sample to the response intensity of a positive control serum contained on that blot. The IgG positive control serum was derived by mixing sera from several individuals with a strong serological response to the 31-kDa antigen group

Serological Statistical Analysis

The serological statistical analyses were conducted using SAS, Version 8. The number of events (gastrointestinal, diarrhea or other) was related to risk factors using a negative binomial distribution (SAS proc GENMOD – NEGBIN distribution). The negative binomial distribution was selected because of over-dispersion of the outcome variables (number of disease events). Two definitions of positivity were considered: 1) Any detectable response and 2) A response greater than or equal to 10% of the positive control. Independent factors in the model included serological response to the *Cryptosporidium* 15/17-kDa, 27-kDa antigens, and the *Giardia* 31-kDa (greater than or equal to 10% of the positive control OD), age group (65-74 years and 75 years and older), sex, high levels of tap water consumption (40 or more glasses per day) and bottled water (8 or more glasses per day). The number of days in town (diary days) was added into each model. Separate models were tested for each of the three groups of illnesses (gastrointestinal, diarrhea and other illnesses). A new episode of illness was presumed after the occurrence of at least one day, 4 days, or 7 days without symptoms. Analyses showed that the choice of the number of days without symptoms (i.e., 1, 4 or 7 or more days) between episodes resulted in only slightly fewer illness episodes but did not affect the relationship between risk factors and outcomes. Therefore, we considered the illness a new event if there was at least 4 days without symptoms. We modeled both the number of illness episodes and number of days of illness.

Stool Samples

Stool samples were collected from subjects in Site A, as this was the area of change in water treatment. Subjects, either adults or children who agreed to participate in this portion of the study, were mailed a stool collection kit and instructions. Once the stool sample was collected, a courier was contacted to pick up the sample and deliver it to a participating hospital clinical laboratory where the samples were split and processed for PCR analysis, ova and parasite microscopic stool examination (O & P) analysis for specimens that were PCR positive, microbiological and viral analysis at US EPA, and storage aliquots. The samples were tested by PCR and O & P for *Cryptosporidia*, *Giardia*, and Microsporidia by staff at Children's Hospital and Regional Medical Center (CHRMC). Samples sent to EPA laboratory at Cincinnati were not analyzed due to the anthrax outbreak. However, the samples were archived. The stool samples that were aliquoted for storage at the hospital were tested for fecal *Cryptosporidia* IgA antibodies using the ELISA protocol (Okhuysen, 1998a). Table 2.4 describes the number of samples collected for this project. The 'Families with children" had both adults and children supplying stool samples, indicated by Children-Adults and Children-Children, respectively.

Table 2.4
Stool samples (site A only)

	Total	Seniors	Children-Adults	Children-Children
Number of samples received in Phase 1	218	59	85	71
Number of samples received in Phase 2	174	51	64	59

Stool PCR Analysis

The methodology used by CHRMC for molecular detection is as follows. One gram of stool specimen was filtered by Fecal Parasite Concentrator. Then, glass wool column filtration was performed to further clean out fecal material from the ova and parasite concentrate before DNA extraction. The concentrated material was then washed by water twice and the final pellet from centrifugation was reconstituted in 0.5 ml buffer for over night Proteinase K digestion. The digested material was then washed by water and reconstituted in water for further physical breakage of ova and parasites by 10 times of freeze-thaw and ultra-sonication. The final pellet material was then used for Fluorescent real-time PCR by Bio-rad iCycler. A set of four reactions were carried out on each specimen: a) GG primers were used for detection of Giardin gene (171bp amplicon) in *Giardia*; b) 18Si primers were used for 18S DNA (298bp) in *C. parvum*; c) PMP primers were used for 18S DNA (250-258bp) in Microsporidia.

Cryptosporidium stool IgA specific ELISA

The methodology used by LCF is as follows. Microtitration plates (Nunc Maxisorp) were coated with 100μl of sonicated *Cryptosporidium* oocysts in carbonate buffer (.2ug of protein per well) for 1 hour at 56°C followed by 16h at 4°C. The plates were washed 3 times with 350μl of wash buffer (phosphate buffer with 0.1% tween20) prior to the addition of a new reagent. The plates were blocked with 200μl of blocking solution (5% dry non fat milk in wash buffer) for 1h at 56°C. 100μl/well of fecal extracts were added at a 1:1 ratio in phosphate buffer for 1h at 56°C. 50μl/well of secondary antibody (Horse Radish Peroxidase (HPR)-goat anti human IgA, Cappel) were added at a 1:1000 ratio in blocking solution for 1h at 56°C. The HRP was activated with 50μl/well of H_2O_2-2,2-azino-di-[3-ethylbenzthiazoline sulfonate] ABTS. Optical densities (OD) were read at 15 and 30 minutes with a Dynex MRX microplate reader at 405nm. All samples were analyzed on the same plate in pairs.

Microtitration plates were rejected and repeated if the positive or negative control had a coefficient of variation (CV) greater than 20%. Individual samples were also repeated if the CV was greater than 20%. An unknown sample was considered positive if the mean OD value was greater than 3 standard deviations above the mean for the negative or conjugate control. An unknown value less than 3 standard deviations above the mean for the negative or conjugate control was considered negative. When an unknown value had discordant results between the two standards, the result was considered indeterminate.

OTHER DATA SOURCES

School Absentee Information

Elementary schools located in Sites A and B were contacted for their absentee records for the fall semesters of 2000 and 2001. Schools that qualified for inclusion were schools whose boundaries were entirely within the service area of either of the surface water distribution systems. Because these schools fell into the appropriate water distribution areas, their students could have qualified for this study. However, it should be noted that students are allowed to attend schools outside of their assigned school districts, so the children in this study may not have attended these schools. Data were not requested from school districts that were served by a combination of both water sources.

School absenteeism data were received from 64 out of the possible 100 elementary schools [5 school districts] for a total of 46,411 students (Table 2.5). The distribution of students by grade is as follows: 120 pre-kindergartners, 5,569 all day kindergartners, 1,439 ½ day kindergartners, 7,670 1st graders, 7,720 2nd graders, 7,985 3rd graders, 7,983 4th graders, and 7,925 5th graders. The majority of school districts that did not submit data were smaller districts of 3 schools or less. Two larger districts with 11 and 13 schools did not submit the requested information due to personnel shortages. The types of information received were diverse, ranging from daily or monthly absentee counts, along with variations in the breakdown of the type of absence [from excused absence to illness]. Also, some of the districts provided student race and ethnicity data.

Table 2.5
Number of schools submitting absenteeism data

Site	Number of schools	Received data
A	25	Yes
A	31	No
B	39	Yes
B	5	No

Data was obtained for two 13-week periods – one in 2000 (weeks beginning 9/4/00 to 11/27/00) and one in 2001 (weeks beginning 9/3/01 to 11/26/01). Four of the schools submitted data for only one phase and were removed from the data set, resulting in 60 schools with information for both phases of the study. Schools were removed from the analysis if data were not available for all grades K-5. This eliminated data from 10 schools, resulting in 50 schools used in the analysis. These schools had 459,553 student-weeks of data. Data were available from 138,860 (30%) student-weeks from Site A and 320,693 student-weeks from Site B. There were 19,586 days of absences in Site A and 52,701 days in Site B.

The unit of analysis was the number of students and number of student absences reported by grade, phase, and site. Therefore for each grade and school, there were 26 entries (one for each week of the study).

For this analysis, a dummy variable was assigned to each school, each week and each grade level (Kindergarten, 1,2,3,4,5). These were used in the regression. The model predicted the number of absences given the school, number of students, the grade level, year and whether the school was located in the intervention or control area. The primary outcome of interest was the parameter for the intervention group (Site A during 2001).

Emergency Department Data

Emergency department visits at a local medical center were tabulated during the study phases. The patients were categorized using the same zip code of residence used in the school absenteeism analysis. They were coded as either diarrhea or control diagnosis. Diarrheal diagnosis included vomiting, diarrhea and or gastroenteritis. The control group included emergency department visit diagnoses for abscess, cellulite, contusion, crush injury, hematoma, sprain, strain, soft tissue injury and fractures.

Water Quality Data

Site A and B's water utility provided water quality data for the corresponding study time frames. The water quality data were collected by the utility at the source and throughout the distribution system for routine monitoring purposes. The data included turbidity, raw water fecal coliform counts, field temperature, pH at the laboratory, specific conductance, total coliform using 15 tubes MTF, chlorine residual, and ultraviolet adsorption at 254 nm. However, for purposes of this study, only turbidity and raw water fecal coliform counts for source water were used.

Fecal coliforms were run on membrane filters. Finished water turbidity was run on a HACH 2100A bench top turbidimeter. Peak day turbidities for compliance were run on HACH 1750C on-line turbidimeters. Average weekly turbidity and coliform counts were calculated for each water supply area (Sites A and B) for each week of the study (both Phase 1 and 2). The daily diary illness reports of illnesses (diarrhea, other gastrointestinal and other illness) occurring during that week were then matched to the mean turbidity or coliform level for the week. Illness event rates in periods with different higher versus low levels of coliform or turbidity were then compared.

CHAPTER 3
RESULTS

GENERAL PARTICIPANT INFORMATION

Families were recruited for Phase 1 of the study as described in Chapter 2. Phase 2 only included families enrolled in Phase 1 who maintained study compliance criteria. Some subjects disenrolled due to moving out of the Site and some could not be located for Phase 2. Subjects were allowed to miss 3 consecutive weeks of Daily Diaries, with attempted follow-up by the field staff via telephone or mailings of the missing forms. If the subjects could be reached, the missing information was taken over the telephone or the forms would be accepted if they were sent to the office late. If the subjects could not be reached, the field staff removed the family from the study. Table 3.1 describes the subjects' study participation. This table also illustrates some of the problems encountered when using all participants, including those who had dropped out. Site A had 181 subjects who were not enrolled in Phase 2, Site B had 99, and Site C had 137. Phase 2 had some subjects who had not participated in Phase 1. These were usually the father of the family was decided to participate because his family had been enrolled during Phase 1.

Table 3.1
General subject participation information

	Site A		Site B		Site C	
	2000	2001	2000	2001	2000	2001
Individuals in Source File	906	900	471	468	444	297
Individuals enrolled but no weekly data, or erroneous id data	14	181	9	101	0	0
Individual ID's corrected or deleted	0	0	0	0	18	0
Individuals not in other year	181	8	99	4	137	8
Net Individuals after above subtractions	711	711	363	363	289	289
Individuals enrolled both years	711		363		289	
Individual Families enrolled both years	277		164		124	
For Individuals enrolled in both years:						
Total Diary Days in Town	72242	102876	33435	48742	23933	41,110
Mean Diary Days in Town per Individual	101.6	144.7	92.1	134.3	82.8	142.2
Mean % of Diary Days in Town & 'Not Well' – per Individual	6.2 %	4.7 %	7.5 %	4.7 %	5.8 %	5.9 %

All further analyses were conducted with only subjects who were enrolled in both Phase 1 and 2. Tables 3.2 and 3.3 provide summaries of participation as measured by the mean number of diary-days, along with the mean number of diary-days enrolled by age, sex, and site for the first and second Phase of the project. The number of diary-days of participation was calculated by excluding days that the participants were out of town. As is apparent from these tables, there was a longer mean duration of enrollment during Phase 2. There was no lag time due to enrollment since all subjects were identified and ready to begin collecting data at the beginning of the study period.

Table 3.2
Mean number of diary-days during phase 1 & 2

Age	Sex	Site	Number Of People	Phase 1 Mean Diary days	Phase 2 Mean Diary days
0-3	Female	A	39	101.5	142.36
		B	16	90.63	139.88
		C	15	58.13	138.67
	Male	A	44	98.77	138.62
		B	21	92.67	126.86
		C	17	82.47	141.18
4-9	Female	A	82	102.93	142.96
		B	37	98.76	130.32
		C	33	82.42	132.70
	Male	A	86	103.91	149.23
		B	32	85.25	139.06
		C	25	75.48	140.80
10-18	Female	A	32	102.72	138.66
		B	12	86.67	118.33
		C	10	87.60	150.20
	Male	A	31	99.42	141.45
		B	11	97.27	94.09
		C	17	87.05	154.82
19-35	Female	A	61	102.89	141.49
		B	23	91.83	135.70
		C	34	78.65	135.09
	Male	A	26	99.54	140.81
		B	10	86.70	139.70
		C	15	77.13	112.40
36-49	Female	A	103	101.28	146.09
		B	59	92.56	136.39
		C	26	78.34	138.00
	Male	A	61	97.52	141.10
		B	29	91.76	124.41
		C	12	86.25	154.58
50-64	Female	A	9	105.89	155.89
		B	14	94.43	131.43
		C	8	103.5	150.63
	Male	A	8	103.25	152.50
		B	7	92.71	125.71
		C	1	112.00	168.00
65 +	Female	A	92	102.40	147.80
		B	58	92.53	147.50
		C	40	94.58	150.95
	Male	A	37	101.16	135.24
		B	33	88.42	137.06
		C	28	884.64	125.11

Table 3.3
Total number of diary-days during phase 1 & 2

Age	Sex	Site	Number Of People	Phase 1 Diary-Days	Phase 2 Diary-Days
0-3	Female	A	39	3962	5552
		B	16	1450	2238
		C	15	857	2080
	Male	A	44	4346	6108
		B	21	1946	2664
		C	17	1402	2400
4-9	Female	A	82	8440	11723
		B	37	3654	4822
		C	33	2720	4379
	Male	A	86	8936	12834
		B	32	2728	4450
		C	25	1887	3520
10-18	Female	A	32	3287	4437
		B	12	1040	1420
		C	10	876	1502
	Male	A	31	3082	4385
		B	11	1070	1035
		C	17	1480	2632
19-35	Female	A	61	6276	8631
		B	23	2112	3121
		C	34	2674	4593
	Male	A	26	2588	3661
		B	10	867	1397
		C	15	1157	1686
36-49	Female	A	103	10432	15047
		B	59	5461	8047
		C	26	2037	3588
	Male	A	61	5949	8607
		B	29	2661	3608
		C	12	1035	1855
50-64	Female	A	9	953	1403
		B	14	1322	1840
		C	8	828	1205
	Male	A	8	826	1220
		B	7	649	8807
		C	1	112	168
65 +	Female	A	92	9421	13598
		B	58	5366	8555
		C	40	3783	6038
	Male	A	37	3743	5004
		B	33	2918	4523
		C	28	2370	3503

SUMMARY OF REPORTED ILLNESS INFORMATION

Using the subjects who were in both Phase 1 and 2, illness episodes were defined using several 'illness definitions' and numbers of symptom free days before a new episode of the same type could occur. Illness definitions used in this analysis include "*Any Illness*", "*Gastrointestinal Illnesses (GI)*", and "*Other Illness*". "*Any Illness*" includes all illnesses recorded by the subject. "*Gastrointestinal Illness*" includes 2 or more episodes of diarrhea, any vomiting, and/or nausea during the day. "*Other Illness*" includes respiratory illnesses and other general illnesses, but not gastrointestinal illnesses. For purposes of this analysis, symptom free days of 7 days, 4 days, and 1 day between episodes were used. Illness rates are reported as a ratio of the number of episodes for a given cause divided by the number of diary days. This is not strictly a rate but other terms are likely to be more misleading.

Tables 3.4 - 3.6 describe the crude illness rates per 100 diary days for "*Any Illness*" events separated by 1 day, 4 days, and 7 days symptom-free for Phases 1 and 2. Tables 3.7 - 3.9 describe the crude illness rates per 100 diary days for "*GI illness*" events separated by 1 day, 4 days, and 7 days symptom-free for Phases 1 and 2. Finally, tables 3.10 - 3.12 describe crude illness rates per 100 diary days for "*Other Illness*" events separated by 1 day, 4 days, and 7 days symptom-free for Phases 1 and 2. These tables show a dramatic decline in disease reporting for most age groups for Phase 2 versus Phase 1.

Table 3.13 provides a summary of any disease reports by phase for a symptom-free period of 1 day. In general, the Phase 2 rate is 70-81% of the Phase 1 rate except for age 50-64 where it was 98% and for 65 and older where it is only 47%. The decline in illness reporting in the elderly may be due to reporting fatigue and may not reflect a decline in illness rates. The decline in illness reporting from Phase 1 to Phase 2 complicates assigning causal factors to changes in illness rates. This problem was approached by comparing the rates of decline between the site populations using multivariate analyses.

Table 3.4
Any illness per 100 diary days - 1 disease-free day between episodes

Age	Sex	Site	ILL-1 Phase 1	ILL-1 Phase 2
0-3	Female	A	2.675	1.675
		B	3.517	1.877
		C	2.511	2.196
	Male	A	2.759	1.965
		B	3.135	2.628
		C	2.376	1.875
4-9	Female	A	1.753	1.382
		B	2.107	1.514
		C	1.624	1.393
	Male	A	1.623	1.270
		B	1.841	1.596
		C	1.398	1.477
10-18	Female	A	1.612	1.104
		B	1.923	0.775
		C	1.834	2.463
	Male	A	1.817	1.049
		B	1.121	0.966
		C	2.002	1.406
19-35	Female	A	2.502	1.715
		B	2.462	1.698
		C	3.076	3.070
	Male	A	1.121	1.338
		B	2.076	2.004
		C	2.759	1.162
36-49	Female	A	2.090	1.542
		B	2.966	1.887
		C	2.114	1.727
	Male	A	1.765	2.035
		B	2.052	1.476
		C	1.736	1.469
50-64	Female	A	1.364	1.617
		B	1.589	0.428
		C	2.295	0.815
	Male	A	0.847	0.738
		B	1.849	1.932
		C	0.000	0.000
65 +	Female	A	1.720	0.757
		B	1.602	0.795
		C	1.462	0.629
	Male	A	1.496	0.829
		B	1.748	0.818
		C	1.025	0.540

Table 3.5
Any illness events per 100 diary days - 4 disease-free days between episodes

Age	Sex	Site	ILL-4 Phase 1	ILL-4 Phase 2
0-3	Female	A	2.145	1.459
		B	2.828	1.743
		C	2.152	2.107
	Male	A	2.299	1.703
		B	2.672	2.252
		C	1.980	1.750
4-9	Female	A	1.505	1.186
		B	1.806	1.410
		C	1.471	1.279
	Male	A	1.477	1.200
		B	1.700	1.438
		C	1.209	1.392
10-18	Female	A	1.521	1.059
		B	1.635	0.563
		C	1.351	2.130
	Male	A	1.557	0.981
		B	0.841	0.966
		C	1.602	1.330
19-35	Female	A	2.199	1.367
		B	2.083	1.346
		C	2.314	2.330
	Male	A	1.005	1.202
		B	1.730	1.575
		C	1.630	0.913
36-49	Female	A	1.658	1.127
		B	2.472	1.535
		C	1.998	1.417
	Male	A	1.446	1.672
		B	1.786	1.290
		C	1.736	1.330
50-64	Female	A	1.154	1.402
		B	1.513	0.356
		C	1.932	0.652
	Male	A	0.605	0.738
		B	1.849	1.705
		C	0.000	0.000
65 +	Female	A	1.444	0.610
		B	1.248	0.690
		C	1.336	0.596
	Male	A	1.042	0.688
		B	1.302	0.774
		C	0.813	0.517

Table 3.6
Any illness events per 100 diary days - 7 disease-free days between episodes

Age	Sex	Site	ILL-7 Phase 1	ILL-7*Phase 2
0-3	Female	A	1.918	1.387
		B	2.552	1.609
		C	2.152	2.017
	Male	A	2.092	1.555
		B	2.518	2.027
		C	1.867	1.500
4-9	Female	A	1.303	1.117
		B	1.423	1.161
		C	1.225	1.119
	Male	A	1.309	1.091
		B	1.558	1.393
		C	1.096	1.364
10-18	Female	A	1.399	1.014
		B	1.538	0.563
		C	1.255	1.931
	Male	A	1.265	0.912
		B	0.748	0.966
		C	1.316	1.216
19-35	Female	A	1.928	1.193
		B	1.657	1.314
		C	1.932	1.894
	Male	A	0.966	1.093
		B	1.269	1.503
		C	1.317	0.747
36-49	Female	A	1.400	1.068
		B	1.941	1.263
		C	1.845	1.230
	Male	A	1.210	1.505
		B	1.634	1.173
		C	1.653	1.275
50-64	Female	A	1.049	1.240
		B	1.362	0.356
		C	1.449	0.652
	Male	A	0.605	0.738
		B	1.695	1.364
		C	0.000	0.000
65 +	Female	A	1.168	0.552
		B	1.006	0.643
		C	1.232	0.580
	Male	A	0.882	0.635
		B	1.062	0.663
		C	0.778	0.517

Table 3.7
Gastrointestinal illness events per 100 diary days - 1 disease-free day between episodes

Age	Sex	Site	GI-1 Phase 1	GI-1 Phase 2
0-3	Female	A	1.666	1.009
		B	2.207	0.894
		C	1.345	1.390
	Male	A	1.610	1.277
		B	1.901	1.426
		C	2.149	1.417
4-9	Female	A	1.031	0.759
		B	1.204	0.892
		C	1.011	0.731
	Male	A	0.795	0.725
		B	1.239	1.079
		C	0.907	0.852
10-18	Female	A	0.791	0.428
		B	0.962	0.141
		C	1.062	1.931
	Male	A	0.973	0.410
		B	0.561	0.193
		C	0.915	0.798
19-35	Female	A	1.625	1.344
		B	1.468	1.250
		C	2.177	2.395
	Male	A	0.618	0.819
		B	1.730	1.217
		C	2.382	1.127
36-49	Female	A	1.256	1.369
		B	2.289	1.094
		C	1.268	1.784
	Male	A	0.840	0.848
		B	1.444	0.776
		C	0.909	1.186
50-64	Female	A	0.944	0.356
		B	1.286	0.543
		C	2.053	0.996
	Male	A	0.121	0.246
		B	1.079	1.364
		C	0.000	0.000
65 +	Female	A	1.253	0.449
		B	1.304	0.713
		C	1.002	0.596
	Male	A	1.122	0.670
		B	1.508	0.641
		C	0.742	0.470

Table 3.8
Gastrointestinal illness events per 100 diary days - 4 disease-free days between episodes

Age	Sex	Site	GI-4 Phase 1	GI-4*Phase 2
0-3	Female	A	1.338	0.774
		B	2.000	0.760
		C	1.076	1.300
	Male	A	1.334	1.146
		B	1.644	1.276
		C	1.640	1.333
4-9	Female	A	0.806	0.648
		B	1.122	0.830
		C	0.950	0.662
	Male	A	0.750	0.678
		B	1.133	0.944
		C	0.718	0.795
10-18	Female	A	0.730	0.428
		B	0.962	0.141
		C	0.965	1.598
	Male	A	0.746	0.388
		B	0.467	0.193
		C	0.801	0.798
19-35	Female	A	1.291	1.054
		B	1.231	0.897
		C	1.769	1.960
	Male	A	0.502	0.738
		B	1.499	1.002
		C	1.505	0.830
36-49	Female	A	1.007	1.070
		B	1.904	0.882
		C	1.191	1.477
	Male	A	0.773	0.732
		B	1.216	0.665
		C	0.909	1.024
50-64	Female	A	0.944	0.285
		B	1.286	0.435
		C	1.691	0.830
	Male	A	0.121	0.246
		B	1.079	1.250
		C	0.000	0.000
65 +	Female	A	1.104	0.427
		B	0.988	0.608
		C	0.877	0.547
	Male	A	0.828	0.529
		B	1.097	0.597
		C	0.601	0.423

Table 3.9
Gastrointestinal illness events per 100 diary days - 7 disease-free days between episodes

Age	Sex	Site	GI-7 Phase 1	GI-7 Phase 2
0-3	Female	A	1.186	0.756
		B	1.793	0.760
		C	1.076	1.210
	Male	A	1.242	1.031
		B	1.490	1.201
		C	1.414	1.250
4-9	Female	A	0.699	0.623
		B	0.958	0.684
		C	0.766	0.639
	Male	A	0.705	0.616
		B	1.027	0.921
		C	0.718	0.795
10-18	Female	A	0.669	0.406
		B	0.962	0.141
		C	0.869	1.531
	Male	A	0.552	0.388
		B	0.467	0.193
		C	0.744	0.798
19-35	Female	A	1.083	0.904
		B	0.994	0.865
		C	1.579	1.524
	Male	A	0.464	0.656
		B	1.153	0.931
		C	1.191	0.771
36-49	Female	A	0.853	0.877
		B	1.428	0.795
		C	1.076	1.226
	Male	A	0.706	0.674
		B	1.102	0.665
		C	0.826	0.916
50-64	Female	A	0.944	0.285
		B	1.210	0.435
		C	1.208	0.747
	Male	A	0.121	0.246
		B	0.924	0.909
		C	0.000	0.000
65 +	Female	A	0.945	0.375
		B	0.783	0.561
		C	0.835	0.530
	Male	A	0.695	0.476
		B	0.925	0.486
		C	0.566	0.423

Table 3.10
Other illness events in per 100 diary days - 1 disease-free day between episodes

Age	Sex	Site	Other-1* Phase 1	Other-1* Phase 2
0-3	Female	A	1.565	1.099
		B	2.000	1.206
		C	1.704	1.300
	Male	A	1.495	0.950
		B	1.696	1.464
		C	0.905	0.708
4-9	Female	A	0.971	0.802
		B	1.149	0.830
		C	0.797	0.708
	Male	A	1.108	0.709
		B	0.673	0.764
		C	0.680	0.852
10-18	Female	A	1.065	0.856
		B	1.250	0.634
		C	1.351	1.132
	Male	A	1.038	0.775
		B	0.654	0.676
		C	1.087	0.722
19-35	Female	A	1.498	0.672
		B	1.563	0.705
		C	1.769	1.568
	Male	A	0.541	0.574
		B	0.692	0.859
		C	0.815	0.474
36-49	Female	A	1.256	0.904
		B	1.483	0.833
		C	0.807	0.892
	Male	A	1.177	0.860
		B	1.064	0.776
		C	0.826	0.647
50-64	Female	A	0.525	0.143
		B	0.605	0.598
		C	0.242	0.415
	Male	A	0.605	0.492
		B	0.770	0.568
		C	0.000	0.000
65 +	Female	A	0.807	0.419
		B	0.466	0.199
		C	0.647	0.182
	Male	A	0.668	0.317
		B	0.377	0.243
		C	0.389	0.117

Table 3.11

Other illness events in per 100 diary days - 4 disease-free days between episodes

Age	Sex	Site	Other-4 Phase 1	Other-4 Phase 2
0-3	Female	A	1.489	1.063
		B	1.793	1.206
		C	1.525	1.255
	Male	A	1.403	0.884
		B	1.593	1.276
		C	0.792	0.708
4-9	Female	A	0.924	0.708
		B	0.958	0.788
		C	0.705	0.708
	Male	A	1.052	0.693
		B	0.673	0.697
		C	0.642	0.852
10-18	Female	A	1.034	0.811
		B	0.962	0.423
		C	1.062	1.065
	Male	A	1.006	0.730
		B	0.654	0.676
		C	1.030	0.684
19-35	Female	A	1.434	0.614
		B	1.373	0.673
		C	1.497	1.415
	Male	A	0.541	0.519
		B	0.577	0.787
		C	0.627	0.474
36-49	Female	A	1.141	0.837
		B	1.300	0.783
		C	0.807	0.725
	Male	A	0.992	0.767
		B	0.988	0.748
		C	0.826	0.647
50-64	Female	A	0.525	0.143
		B	0.605	0.489
		C	0.242	0.415
	Male	A	0.484	0.492
		B	0.770	0.455
		C	0.000	0.000
65 +	Female	A	0.658	0.287
		B	0.429	0.187
		C	0.626	0.182
	Male	A	0.534	0.317
		B	0.343	0.243
		C	0.247	0.117

Table 3.12
Other illness events in per 100 diary days - 7 disease-free days between episodes

Age	Sex	Site	Other-7* Phase 1	Other-7* Phase 2
0-3	Female	A	1.489	1.009
		B	1.724	1.117
		C	1.525	1.255
	Male	A	1.288	0.835
		B	1.593	1.164
		C	0.792	0.667
4-9	Female	A	0.853	0.699
		B	0.821	0.747
		C	0.613	0.617
	Male	A	0.974	0.662
		B	0.673	0.674
		C	0.605	0.852
10-18	Female	A	1.004	0.789
		B	0.865	0.423
		C	0.965	0.932
	Male	A	0.909	0.707
		B	0.561	0.676
		C	0.858	0.684
19-35	Female	A	1.307	0.602
		B	1.231	0.641
		C	1.279	1.285
	Male	A	0.541	0.492
		B	0.461	0.787
		C	0.564	0.474
36-49	Female	A	1.045	0.784
		B	1.172	0.696
		C	0.769	0.669
	Male	A	0.857	0.697
		B	0.988	0.721
		C	0.826	0.593
50-64	Female	A	0.525	0.143
		B	0.530	0.435
		C	0.242	0.332
	Male	A	0.484	0.492
		B	0.770	0.455
		C	0.000	0.000
65 +	Female	A	0.520	0.279
		B	0.391	0.187
		C	0.585	0.149
	Male	A	0.401	0.317
		B	0.308	0.243
		C	0.247	0.117

Table 3.13
Phase 2 rate of any illness with one illness-free day between episodes as a percent of phase 1 (both sexes, all areas)

Age group	Phase 2 Illness Rate as a Percent of Phase 1 Rate
0-3 years	70% of Phase 1 Rate
4-9 years	81% of Phase 1 Rate
10-18 years	72% of Phase 1 Rate
19-35 years	80% of Phase 1 Rate
36-49 years	79% of Phase 1 Rate
50-64 years	98% of Phase 1 Rate
65+ years	47% of Phase 1 Rate

COMPARISON OF DEMOGRAPHIC FACTORS WITH ILLNESS RATES

The demographic, enrollment, and water consumption data were related to the illness diary reports for each period. The number of diarrheal, other gastrointestinal and other illness days and episodes were examined in relation to the demographic, enrollment and water consumption data using the multi-level hierarchical analysis. The variables selected, based on findings from prior studies, were age group (0-9 years, 10-19 years 65-74 years and 75 years and older versus ages 35-64 years, male gender, having a dog or cat in the household, completing some college education, white race, and consuming high levels of tap water in the home, work or elsewhere (40 or more glasses per day).

Table 3.14 shows the results of data analysis for both phases combined. For illnesses of any duration, age, gender and year were predictive of illness rates for each illness category. Males reported fewer gastrointestinal, highly credible gastrointestinal events and other events. Site A had fewer diarrhea events than Site B. In general, rates of reported illness declined in the second period of the study as mentioned earlier. Of interest is that Site A in the second year of the study had a higher rate of diarrhea and highly credible gastrointestinal events than Site B.

Table 3.15 reports the same information except events were considered to have occurred only if they persisted for at least 2 days. Again, Site A compared to Site B in Phase 2 had higher rates of diarrhea and highly credible gastrointestinal events.

Table 3.14
Relationship between illness rates and age, gender, phase and site
(illnesses of any duration)

Variable	GI p-value	Diarrhea p-value	HCGE* p-value	Other p-value
Age 0-9 vs. ref	(-) 0.007	(+) 0.01	(+) 0.47	(+) 0.08
Age 10-19 vs. ref	(-) 0.04	(-) 0.004	(-) 0.005	(-) 0.15
Age 20-35 vs. ref	(-) 0.22	(-) 0.0001	(-) 0.0007	(-) 0.18
Age 35-49 (ref)	-----	-----	-----	----
Age 50-64 vs. ref	(-) 0.22	(+) 0.78	(-) 0.21	(-) 0.05
Age 65-74 vs. ref	(-) 0.006	(-) 0.48	(-) 0.10	(-) 0.0001
Age 75+ vs. ref	(-) 0.0002	(-) 0.003	(-) 0.0005	(-) 0.0001
Male vs. Female gender	(-) 0.001	(-) 0.06	(-) 0.003	(-) 0.0001
Phase 2 vs. Phase 1	(-) 0.0001	(-) 0.001	(-) 0.0001	(-) 0.002
Site A in Phase 1 vs. Site B in Phase 1	(-) 0.14	(-) 0.02	(-) 0.03	(+) 0.59
Site C in Phase 1 vs. Site B in Phase 1	(-) 0.91	(-) 0.74	(+) 0.62	(-) 0.42
Site A in Phase 2 vs. Site B in Phase 2	(+) 0.13	(+) 0.07	(+) 0.01	(-) 0.64
Site C in Phase 2 vs. Site B in Phase 2	(+) 0.04	(+) 0.03	(+) 0.04	(+) 0.26

Diarr - reporting a high number of diarrheal illnesses on the intake questionnaire
HCGE – Highly credible gastrointestinal events (Payment (1991a, 1991b)
Direction of the effect (+ or -)

Table 3.15
Relationship between illness rates and age, gender, phase and site
(illnesses lasting 2 or more days)

Variable	GI p-value	Diarrhea p-value	HCGE* p-value	Other p-value
Age 0-9 vs. ref	(-) 0.0002	(+) 0.42	(-) 0.37	(+) 0.002
Age 10-19 vs. ref	(-) 0.03	(-) 0.0002	(-) 0.0002	(-) 0.34
Age 20-35 vs. ref	(-) 0.02	(-) 0.0001	(-) 0.0005	(-) 0.22
Age 35-49 (ref)	-----	-----	-----	-----
Age 50-64 vs. ref	(-) 0.60	(-) 0.87	(-) 0.45	(-) 0.25
Age 65-74 vs. ref	(-) 0.11	(-) 0.24	(-) 0.21	(-) 0.0001
Age 75+ vs. ref	(-) 0.002	(-) 0.009	(-) 0.001	(-) 0.0001
Male vs. Female gender	(-) 0.0002	(-) 0.14	(-) 0.01	(-) 0.007
Phase 2 vs. Phase 1	(-) 0.01	(-) 0.0001	(-) 0.0001	(-) 0.008
Site A in Phase 1 vs. Site B in Phase 1	(-) 0.20	(-) 0.002	(-) 0.07	(-) 0.78
Site C in Phase 1 vs. Site B in Phase 1	(+) 0.78	(-) 0.43	(+) 0.38	(-) 0.04
Site A in Phase 2 vs. Site B in Phase 2	(+) 0.18	(+) 0.01	(+) 0.004	(-) 0.75
Site C in Phase 2 vs. Site B in Phase 2	(+) 0.34	(+) 0.08	(+) 0.16	(+) 0.15

Diarr - reporting a high number of diarrheal illnesses on the intake questionnaire
HCGE – Highly credible gastrointestinal events (Payment 1991a, 1991b)
Direction of the effect (+ or -)

COMPARISON OF WATER DATA WITH ILLNESS RATES

The purpose of looking at raw water data is to determine if illnesses rates were related to turbidity and fecal coliform levels. Average weekly turbidity and coliform levels were calculated for each surface water supply area (Sites A and B) for each week of the study (both Phases 1 and 2. The turbidity and fecal coliform levels were for the source water and not for the distribution system. The daily diary illness reports of illnesses (diarrhea, other gastrointestinal and other illness) occurring during that week were then assigned the mean turbidity or fecal coliform level for the week. The results for Phase 1 and 2 were then combined and summarized in Tables 3.16 and 3.17. The data are displayed for each age group (0 to 3 years, 4 to 9 years, 10 to 18 years, 19 to 35 years, 36-64 years and 65 years and older. No delay or latency was considered for this pilot work. The objective of the analysis was to determine if there was any suggestive evidence that raw water turbidity or coliform level was related to illness incidence as determined by the daily diaries.

With the exception of other illnesses in the 36 to 49 year old group, illnesses did not occur more commonly at times of high turbidity. In the 36 to 49 year age group, it appeared that more "*other illness*" (not diarrhea or other gastrointestinal illness) occurred when the turbidity was highest. This is a crude analysis relating the source water turbidity and fecal coliform levels with illness reports. However, compared to prior analyses, it is likely that the daily diary reports better characterize the temporal sequence of illness in the population. There was a statistical relationship between turbidity and illness in the 36 to 49 year old age group. However, in this case, the highest counts of diarrhea or other gastrointestinal illnesses did not occur during the periods of highest turbidity. We did not conduct multiple statistical models attempting to relate source water fecal coliform or turbidity levels and it is possible that, with some combination of latency, time period and age group, one may find a relationship that may appear to be statistically significant if the relationship was expected prior to the study. However, it is also reasonable to assume that if these markers of water quality are causally related to disease occurrence, there should, at least, be some indication of a higher illness rate in periods of high levels of these markers.

Table 3.16
Comparison of source water fecal coliform levels with illness rates at sites A and B

Age 0 to 3 years	Fecal Coliform Levels			
	≤ 2	2 to ≤ 5	5 to ≤ 10	> 10
Diarrhea	19	105	148	104
Other Gastrointestinal	8	38	47	33
Other Illness	21	120	160	114
Chi-square 0.92 (6 d.f.)	p=0.98			
Age 4 to 9 years	**Fecal Coliform Levels**			
	≤ 2	2 to ≤ 5	5 to ≤ 10	> 10
Diarrhea	10	95	165	112
Other Gastrointestinal	6	84	100	79
Other Illness	18	177	209	159
Chi-square 7.03 (6 d.f.)	p=0.32			
Age 10 to 18 years	**Fecal Coliform Levels**			
	≤ 2	2 to ≤ 5	5 to ≤ 10	> 10
Diarrhea	3	26	43	26
Other Gastrointestinal	3	34	45	37
Other Illness	9	61	91	59
Chi-square 1.78 (6 d.f.)	p=0.94			
Age 19 to 35 years	**Fecal Coliform Levels**			
	≤ 2	2 to ≤ 5	5 to ≤ 10	> 10
Diarrhea	12	86	154	136
Other Gastrointestinal	8	67	78	67
Other Illness	12	113	139	99
Chi-square 10.94 (6 d.f.)	p=0.09			
Age 36 to 49 years	**Fecal Coliform Levels**			
	≤ 2	2 to ≤ 5	5 to ≤ 10	> 10
Diarrhea	24	142	250	164
Other Gastrointestinal	13	93	150	88
Other Illness	29	187	236	166
Chi-square 7.09 (6 d.f.)	p=0.31			
Age 50 to 64 years	**Fecal Coliform Levels**			
	≤ 2	2 to ≤ 5	5 to ≤ 10	> 10
Diarrhea	1	15	33	23
Other Gastrointestinal	0	5	16	9
Other Illness	2	16	20	12
Insufficient expected counts				
Age 65+ years	**Fecal Coliform Levels**			
	≤ 2	2 to ≤ 5	5 to ≤ 10	> 10
Diarrhea	12	105	155	115
Other Gastrointestinal	8	68	89	72
Other Illness	4	84	103	51
Chi-square 10.33 (6 d.f.)	p=0.112			

d.f. – degrees of freedom

Table 3.17
Comparison of source water turbidity levels with illness rates at sites A and B

Age 0 to 3 years	Turbidity Levels (NTU's)				
	≤ 0.3	0.3 to ≤ 0.4	0.4 to ≤ 0.5	0.5 to ≤ 0.6	> 0.6
Diarrhea	71	159	50	33	63
Other Gastrointestinal	21	53	14	16	22
Other Illness	73	150	55	50	93
Chi-square 8.73 (8 d.f.)	p=0.36				
Age 4 to 9 years	**Turbidity Levels (NTU's)**				
	≤ 0.3	0.3 to ≤ 0.4	0.4 ≤ 0.5	0.5 to ≤ 0.6	> 0.6
Diarrhea	73	171	43	40	55
Other Gastrointestinal	60	104	40	30	35
Other Illness	97	224	78	59	105
Chi-square 10.8 (8 d.f.)	p=0.21				
Age 10 to 18 years	**Turbidity Levels (NTU's)**				
	≤ 0.3	0.3 to ≤ 0.4	0.4 ≤ 0.5	0.5 to ≤ 0.6	> 0.6
Diarrhea	22	29	15	7	25
Other Gastrointestinal	33	37	15	12	22
Other Illness	38	86	21	27	48
Chi-square 11.4 (8 d.f.)	p=0.18				
Age 19 to 35 years	**Turbidity Levels (NTU's)**				
	≤ 0.3	0.3 to ≤ 0.4	0.4 ≤ 0.5	0.5 to ≤ 0.6	> 0.6
Diarrhea	66	175	39	36	72
Other Gastrointestinal	38	94	20	28	40
Other Illness	72	153	40	39	59
Chi-square 4.1 (8 d.f.)	p=0.85				
Age 36 to 49 years	**Turbidity Levels (NTU's)**				
	≤ 0.3	0.3 to ≤ 0.4	0.4 ≤ 0.5	0.5 to ≤ 0.6	> 0.6
Diarrhea	108	264	58	63	87
Other Gastrointestinal	81	126	46	33	58
Other Illness	119	221	80	74	124
Chi-square 19.7 (8 d.f.)	p=0.01				
Age 50 to 64 years	**Turbidity Levels (NTU's)**				
	≤ 0.3	0.3 to ≤ 0.4	0.4 ≤ 0.5	0.5 to ≤ 0.6	> 0.6
Diarrhea	23	29	6	6	8
Other Gastrointestinal	9	11	4	4	2
Other Illness	8	18	10	5	9
Insufficient expected counts					
Age 65+ years	**Turbidity Levels (NTU's)**				
	≤ 0.3	0.3 to ≤ 0.4	0.4 ≤ 0.5	0.5 to ≤ 0.6	> 0.6
Diarrhea	110	160	41	32	54
Other Gastrointestinal	74	97	21	10	35
Other Illness	68	84	27	20	43
Chi-square 7.87 (8 d.f.)	p=0.44				

d.f. – degrees of freedom

SEROLOGICAL ANALYSIS

In Phase 1, 262 sera (124 from Site A, 65 from Site B and 73 from Site C) and in Phase 2, 260 sera (125 from Site A, 39 from Site B and 96 from Site C) were obtained from subjects who participated in both phases of the study. It should be noted that the subjects participated in the diary portion of the study in both phases, not necessarily the blood sample portion in both phases. Initial analysis looked at the presence or absence of a detectable band, with further analysis using an OD value of 10% or greater of the positive control. Table 3.18 details the response levels to the *Cryptosporidium* 15/17-kDA and 27-kDa markers and the Giardia 31-kDa marker during Phase 1 of the study. 51% of the serological samples had a detectable response to the 15/17-kDa antigen group, 69% to the 27-kDa group, and 84% to 31-kDa *Giardia* antigen in Phase 1.

As shown in Table 3.19, 50% of the serological samples had a detectable response to the 15/17-kDa group, 75% to the 27-kDa group, and 82% to the 31-kDa group in Phase 2. Considering a positive response as 10% of the positive control or greater, then of the Phase 1 serological samples, 42% were positive for the 17-kDa marker, 57% for the 27-kDa marker and 44% for the 31-kDa *Giardia* marker. In Phase 2, 43% of the serological samples were positive for the 15/17-kDa marker, 62% were positive for the 27-kDa marker, and 48% were positive for the 31-kDa *Giardia* marker.

There was a higher prevalence of responses to the 31-kDa *Giardia* antigen in Phase 1 in the intervention site (p=0.03) but not in Phase 2 (p=0.16). There was a significantly higher prevalence of responses greater than 10% of the positive control to the 15/17-kDa and 27-kDa *Cryptosporidium* antigen groups in the groundwater community during Phase 2 (p=0.05 and p=0.001, respectively) and suggestive evidence for a higher prevalence of responses to the *Giardia* antigen (p=0.08).

Table 3.18
Phase 1 – Intensity of *Cryptosporidium* and *Giardia* responses by area as a percent of positive control

Area	All Areas Combined	Site A (Unfiltered surface water with chlorine disinfection)	Site B (unfiltered surface water with chlorine disinfection)	Site C (Groundwater with chlorine disinfection)	p-value (Chi squared test by area)
15/17-kDa >0	51%	54%	52%	46%	0.48
27-kDa >0	69%	71%	65%	70%	0.65
31-kDa >0	84%	90%	75%	82%	0.03
17-kDa ≥10%	42%	48%	40%	34%	0.14
27-kDa ≥10%	57%	61%	52%	53%	0.39
31-kDa ≥10%	44%	49%	35%	56%	0.19
N=		124	65	73	

Table 3.19
Phase 2 – Intensity of *Cryptosporidium* and *Giardia* responses by area as a percent of the positive control

Area	All Areas Combined	Site A (Filtered, surface water with ozone and chlorine disinfection)	Site B (unfiltered surface water with chlorine disinfection)	Site C (Groundwater with chlorine disinfection)	p-value (Chi squared test by area)
15/17-kDa >0	50%	47%	41%	57%	0.16
27-kDa >0	75%	68%	69%	88%	0.02
31-kDa >0	82%	85%	71%	83%	0.16
17-kDa ≥10%	43%	42%	28%	51%	0.05
27-kDa ≥10%	62%	50%	54%	80%	0.001
31-kDa ≥10%	48%	55%	41%	42%	0.08
N=		125	39	96	

Comparison of Serological and Other Factors to Illness Episodes and Days

Completed daily diaries, water consumption surveys, and sera were available from 247 participants who participated in both Phase 1 and 2. The number of gastrointestinal, diarrheal, and other health events and the number of days of illness were related to risk factors using a multivariate negative binomial regression model (Table 3.20, Table 3.21). In this analysis, a serological response was defined as being 10% of the positive control or greater. Older age was related to a lower risk of gastrointestinal and illness events. In Phase 2, the surface water sites had a lower rate of illness events than the ground water site. There was suggestive evidence that high levels of tap water consumption were associated with a lower risk of diarrheal events and days of illness during Phase 2.

During Phase 2, statistically significant protective relationships were observed between serological responses of 10% or greater to the 15/17-kDa antigen and diarrheal and gastrointestinal events (Table 3.21). A protective relationship between serological responses to *Cryptosporidium* antigens and other illness (colds, chills, etc) was not observed, however, a weak protective relationship was observed for to the 31-kDa *Giardia* antigen. The statistical significance of this relationship disappeared if the number of disease-free days were changed to either one or seven days.

Table 3.20
Phase 1 – Statistical significance of relationship between illness episodes or days and age, serology, gender and risk factors (p-value)

Variable	#Diarrhea Events	# Diarrhea Days	# GI Events	# GI Days	# Other Events	# Other Days
17-kDa 10%.	N.S.	N.S	N.S.	N.S.	N.S.	Ns
27-kDa 10%	N.S.	N.S.	N.S.	N.S.	N.S.	Ns
31-kDa 10%	N.S.	N.S.	N.S.	N.S.	N.S.	(+) 0.04
Age 65-74	N.S.	N.S.	(-) 0.005	(-) 0.004	(-) 0.0002	(-) 0.005
Age 75+	N.S.	N.S.	N.S.	N.S.	(-) 0.0005	(-) 0.0001
Male Gender	N.S.	N.S.	0.008	(-) 0.004	N.S.	Ns
Site A	N.S.	N.S.	N.S.	N.S.	N.S.	N.S.
Site B	N.S.	N.S.	N.S.	N.S.	N.S.	N.S.
High Other Tap Water Use	N.S.	N.S.	N.S.	N.S.		(+) 0.06
High Home Tap Water Use	N.S.	N.S.	N.S.	(-) 0.05	N.S.	N.S.
Dog in household	N.S.	N.S.	(+) 0.01	N.S.	N.S.	N.S.
Cat in household	N.S.	N.S.	N.S.	N.S.	N.S.	N.S.

The statistical model was a negative binomial regression.
N.S. – not significant
+/- indicator of direction of effect

Table 3.21
Phase 2 – Statistical significance of relationship between illness episodes or days and age, serology, gender and risk factors (p-value)

Variable	#Diarrhea Events	# Diarrhea Days	# GI Events	# GI Days	# Other Events	# Other Days
17-kDa 10%.	(-) 0.07	(-) 0.002	N.S.	(-) 0.05	N.S.	N.S.
27-kDa 10%	N.S.	(-) 0.09.	N.S.	N.S.	N.S.	N.S.
31-kDa 10%	N.S.	N.S.	N.S.	(-) 0.04	N.S.	N.S.
Age 65-74	(-) 0.002.	(-) 0.001	(-) 0.007	(-) 0.0003	(-) 0.01.	(-) 0.005
Age 75+	(-) 0.06.	(-) 0.03	N.S.	N.S.	(-) 0.0001	(-) 0.0001
Male Gender	N.S.	N.S.	(-) 0.05	(-) 0.03	N.S.	Ns
Site A	(-) 0.01.	(-) 0.05	N.S.	N.S.	(+) 0.07	(-) 0.05
Site B	(-) 0.03.	N.S.	N.S.	N.S.	N.S.	Ns
High Other Tap Water Use	N.S.	N.S.	N.S.	N.S.	N.S.	(-) 0.08
High Home Tap Water Use	(-) 0.03	(-) 0.05	N.S.	N.S.	N.S.	N.S.
Dog in household	N.S.	N.S.	N.S.	N.S.	N.S.	N.S.
Cat in household	N.S.	N.S.	N.S.	N.S.	N.S.	N.S.

The statistical model was a negative binomial regression.
N.S. – Not Significant
+/- indicator of direction of effect

STOOL RESULTS

Summary of Stool PCR Analysis

Following the certification of the methodology, 393 stool specimens were analyzed by real time PCR, using appropriate positive control stools from *Giardia* ATCC50137, Giardia genomic DNA ATCC30888D, *Encephalitozoon hellum* ATCC:CDC0291, *Cryptosporidium* oocysts from NIH AIDS Reagent and Reference Reagent Program:GCH-1 and Mason Farm, Idaho. The sensitivities of the technique were 30 copies for Giardia, 100 copies for *Cryptosporidium* oocysts, and 200 copies for *Encephalitozoon hellum* spores respectively. Stools were frozen after receipt, and thawed and DNA extracted by the procedure described in the Procedure Manual. Of these 393 subject stools, 2 (0.5%) were positive for *Giardia lamblia (*Giadin gene), and none (0%) for both *Cryptosporidium parvum* and *Encephalitozoon hellum* sequences. Viability of the organisms from which the signals were derived could not be determined. Ova and parasite examination of the specimens #8 and #122 demonstrated no *Giardia*-like organisms in the specimen. Information from pilot studies using type strains of the parasite cysts has determined that physical breakage procedures by freeze-thaw and ultra-sonication is very important to extract cyst DNA. However, the DNA from the cyst eggs tent stay with the broken shells. Therefore, DNA extraction by filtration methods may compromise the DNA recovery.

From the data, it can be concluded that real time PCR is more sensitive than microscopy, yet its specificity as applied to this population is not determinable. We also conclude that the rates of asymptomatic giardiasis or cryptosporidiosis are low or negligible in this population.

Summary of *Cryptosporidium* IgA ELISA Using Stool Samples

In Phase 1, 183 samples were analyzed for fecal IgA antibody response to *Cryptosporidium*. 43 samples were considered positive by the previously described definition. 118 samples were considered to be negative and 22 had indeterminate results. In Phase 2, 160 samples were analyzed. Thirty-one of these samples were considered positive. One hundred and nine samples were considered negative and 20 had indeterminate results. Table 3.22 further describes the stool results.

Table 3.22
Summary of stool sample *Cryptosporidium* IgA antibody analysis

Result	Phase 1	Number	Percent total	Phase 2	Number	Percent total
Positive		43	23.5%		31	19.4%
Negative		118	64.5%		109	68.1%
Indeterminate		22	12.0%		20	12.5%
Total		183	100%		160	100%

OTHER DATA ANALYSES

Analysis of School Absenteeism Data

School absence data were obtained for two 13-week periods – one in 2000 (weeks beginning 9/4/00 to 11/27/00) and one in 2001 (weeks beginning 9/3/01 to 11/26/01) from Site A and Site B. Schools were removed from the database if data were not available for grades K-5 for both time periods. This eliminated data from 14 schools, resulting in information from 50 schools. These schools had 459,553 student-weeks of data. Data were available from 138,860 (30%) student-weeks from Site A and 320,693 student-weeks from Site B. There were 19,586 days of absences in Site A and 52,701 days in Site B. The unit of analysis was the number of students and number of student absences summarized by grade, phase, and site.

School absence data were analyzed as categorical data with discrete counts (Table 3.23). The number of days absent was tallied by grade level, time period, and area of the city (intervention versus control). These counts were analyzed using SAS proc genmod. Although the analysis followed the outline of a negative binomial model, to reduce over-dispersion, we used the negative binomial distribution rather than the Poisson distribution. This resulted in a ratio of the deviance to the degrees of freedom close to one (1.61), resulting in valid estimates of the statistical significance.

The findings show that Site A had minimal change from Phase 1 to Phase 2. Of the 6 grades and 2 sites, which resulted in 12 comparisons, 8 increased in absenteeism, 3 were negative, and 1 had no change from Phase 1 to Phase 2. However, this is a crude measure of effect and could be subject to many other factors.

The findings also show that lower grade levels (kindergarten, first, second and third grade) had higher rates of school absences. The second phase had a lower rate of absence and the intervention area had a lower rate of school absenteeism. The intervention area during the second phase of the study (i.e. the time of the intervention) was statistically lower than expected with a p-value of 0.04 (Table 3.24).

Table 3.23
Number of days absent and in school by grade, phase, and site

Grade Level	Phase	Site	Days Absent	Days in School	% Days Absent
0	1	A	2043	11683	17.5%
		B	7163	24624	29.1%
	2	A	1995	11375	17.5%
		B	7904	23343	33.9%
1	1	A	1684	11719	14.4%
		B	8349	26863	31.1%
	2	A	1868	11975	15.6%
		B	9123	26549	34.4%
2	1	A	1688	12678	13.3%
		B	7433	26847	27.7%
	2	A	1633	11576	14.1%
		B	8004	26392	30.3%
3	1	A	1481	12373	12.0%
		B	7499	28531	26.3%
	2	A	1459	12619	11.6%
		B	7728	26898	28.7%
4	1	A	1444	12633	11.4%
		B	7484	28479	26.3%
	2	A	1214	11000	11.0%
		B	7597	27241	27.9%
5	1	A	1428	12455	11.5%
		B	6732	27044	24.9%
	2	A	1453	12713	11.4%
		B	7831	27878	28.1%

Table 3.24
p-values of grade level, site, phase, and intervention

Variable	p-value
Grade 0	<0.0001
Grade 1	<0.0001
Grade 2	0.0004
Grade 3	0.36
Grade 4	0.87
Grade 5	--
Phase 2	0.0004
Site A	<0.0001
Intervention	0.04

Analysis of Emergency Department Data

Patients who were seen at the emergency department from June 1, 2000 through December 31, 2000, and from June 1, 2001 through December 31, 2001 were sorted by their zip code of residence into Site A and Site B. Patients who did not fit into the recognized zip codes were coded as "other". The "other" category also included subjects with missing information. The "Diarrheal" group included diagnoses of vomiting, diarrhea, and or gastroenteritis. The "Control" group included emergency department visit diagnoses for abscess, cellulite, contusion, crush injury, hematoma, sprain, strain, soft tissue injury, and fractures. Tables 3.25 and 3.26 describe the number of visits during Phase 1 and Phase 2.

Table 3.25
Phase 1 emergency department visits

	Diarrhea		Controls	
	Number	Percent	Number	Percent
Total Patients	1355	100	2801	100
Site A	181	13.4	378	13.5
Site B	365	26.9	699	25
Other	809	59.7	1724	61.5

There were no differences between Site A and Site B when comparing the number of diarrhea and control group visits during Phase 1 (p=0.4).

Table 3.26
Phase 2 emergency department visits

	Diarrhea		Controls	
	Number	Percent	Number	Percent
Total Patients	1135	100	2367	100
Site A	150	13.2	334	14.1
Site B	286	25.2	550	23.2
Other	699	61.6	1483	62.7

There were no differences between Site A and Site B when comparing the diarrhea and control groups during Phase 2 (p=0.2). When comparing the population as a whole, there was no difference in the number of visits for diarrhea in Site A versus Site B when comparing Phase 1 to Phase 2 (p=0.7.) Comparison of the emergency department data with the diary data was not done.

CHAPTER 4
DISCUSSION

The primary objective of the study was to estimate drinking-water-attributable endemic levels of gastrointestinal diseases. This was done by measuring changes in waterborne disease occurrence associated with changes in water treatment. A secondary objective was to determine if drinking-water-attributable endemic levels of gastrointestinal diseases are elevated in a community that uses unfiltered water from a well-protected surface water source. This was done by comparing illness rates between the intervention site before and after the intervention and by comparing illness rates between the intervention site and the two control sites prior to the intervention.

Since we did not observe any reduction in disease occurrence between phases, it would appear that the drinking water attributable disease risks for this population was zero. This would also be the case for the drinking water attributable disease risks due to *Cryptosporidium*. This conclusion is, of course, subject to the limitations of the study design and sample size. One of the primary design limitations is that people were enrolled had previously lived in the city. If we had recruited people who recently moved into the city, the findings may have been different. The limitation of most waterborne disease intervention studies, including this one, is that protective immunity was not considered. If, instead of recruiting residents of at least six months, we had focused on people without prior exposure to the water supply, then a different conclusion may have been reached.

For example, a cryptosporidiosis outbreak in Talent, Oregon in 1992 and another in Collingwood, Ontario, in1996 disproportionately caused illness in visitors rather than in residents (Frost et al. 2000, Frost et al. 1998a, Frost et al. 1998b). In fact, in Collingwood, more than half of all laboratory confirmed cases of cryptosporidiosis occurred in visitors (Frost el al. 2000), prompting physicians to argue in a local newspaper that no waterborne disease outbreak could be occurring in Collingwood. Similarly, in Talent, Oregon, a high attack rate of cryptosporidiosis occurred in out-of-town visitors attending a wedding celebration. A local tanker truck driver noticed the low cryptosporidiosis attack rate in Talent residents and decided to truck Talent water to a Medford nursing home. A high cryptosporidiosis attack rate followed the delivery of the Talent drinking water (Frost et al. 1998b).

These outbreaks and others suggested that protective immunity might result from repeated low-level pathogen exposure, thereby reducing the risk of illness from later high levels of exposure (Frost et al. 1997). If correct, the population at risk may not be long-term residents but people new to the community or visiting. These people often lack protective immunity to agents in the local water and may be a high risk of illnesses.

Public health officials have assumed that a reduction in exposure to pathogens will result in a reduction in gastrointestinal and/or diarrheal illness events. This is based on an assumption that exposure to a pathogen can increase the risk of both infection and disease. However, the risks of infection may differ from the risks of disease because disease risks may be related to the number of prior infections. Although it is difficult to prove, it is possible that low-dose exposures can result in an immunizing infection that does not result in disease. It is also possible that continued low-dose exposures can maintain protection from illness.

The increased occurrence of diarrheal illness in Site A during Phase 2 was unexpected. This could have been due to the occurrence of an unmeasured change in either risks or exposures. We did observe a small decline in serological responses from Phase 1 to Phase 2 in

Sites A and B. We also observed statistically significant protective effects for diarrheal and gastrointestinal illnesses in people with a serological response to the 15/17-kDa antigen of at least 20% of the positive control. Since these protective effects were observed equally in all three areas, we believe that drinking water was unlikely to be the source of infection and also was unlikely to be the cause of the increased illness occurrence during Phase 2 in Site A.

It is especially important for the assumed benefits from water treatment investments that are capable of removing or inactivating all or almost all pathogens in our drinking water to be validated. Serological studies, such as the one reported here, can increase the understanding of the underlying causes of recent giardiasis and cryptosporidiosis outbreaks. It is even possible that the recent emergence of *Cryptosporidium* and *Giardia* as common waterborne pathogens has resulted from temporal changes in endemic levels of infection and from reduced protective immunity.

It is possible that current efforts to study waterborne diseases have focused on study methodology, such as the need for randomization, rather than on improving our understanding the causes of waterborne disease. Depending on the mechanisms by which agents are transmitted and the population's protective immunity, if any, it may be possible to better allocate resources to minimize the risks of waterborne diseases. It is possible that many of the intervention studies, including this one, are unlikely to provide the needed information to explain why new pathogens have emerged in the past or to predict which organisms will emerge as pathogens in the future. Furthermore, it is critical that the regulatory agencies be better able to predict whether the mandated interventions are effective in reducing disease risks, have no effect or maybe even increase the risks.

LIMITATIONS

As with any study, there are limitations that result from the study design and the analysis methods used. People or families who enrolled in the study may differ from people who did not enroll. Although we restricted the study to individuals enrolled in both study phases, the individuals who were not enrolled in both phases may have been different from those who remained enrolled. Therefore, the study population may not be comparable to either the general population or the group originally recruited. Although there is good reason to believe that this is true, it remains unclear what, if any, effect this would have on the study findings. Finally, if protective immunity plays an important role in reducing the burden of disease, then prior exposure to the water supply, which occurred in all individuals, may have reduce the burden of waterborne disease in this population.

Although the analyses made efforts to adjust for risk factors, it is possible that other factors may have been important in either causing differences in disease occurrence between the areas or in masking true differences in disease occurrence. However, this would be true for any study, given the limitations in our knowledge of causes of diarrheal disease and limited information of other exposures in the study population.

CHAPTER 5
CONCLUSIONS AND RECOMMENDATIONS

This study had two objectives. The first was to estimate drinking-water-attributable endemic levels of gastrointestinal diseases. Because we did not see a reduction in illness rates, the estimated endemic disease risk was zero. This conclusion is, however, subject to a variety of limitations resulting from the study design. This would also apply to the determination of the drinking water attributable endemic disease level due to *Cryptosporidium*. In particular, if protective immunity were a factor in reducing the level of risk of waterborne disease, then the selection of individuals with prior residence in the area would reduce their risk of waterborne disease. Therefore, they may not benefit from improvements in drinking water treatment.

A secondary objective was to determine if drinking-water-attributable endemic levels of gastrointestinal diseases are elevated in a community that uses unfiltered water from well-protected surface water. Site A had a higher rate of reported diarrhea and highly credible gastrointestinal events than Site B in Phase 2. However, overall, the rates of reported illnesses declined in the second period of the study. We did not observe higher levels of *Cryptosporidium* serological responses for users of surface water versus ground water in this study. This may have been due to the high quality of the source water. The two watersheds supplying surface drinking water were well protected, receiving no human or domestic animal fecal waste.

People with increased serological response to the 15/17- and 27- kDa *Cryptosporidium* antigen groups had a reduced risk of diarrheal and gastrointestinal illness. Protective relationships were not observed between the occurrence of other health events (colds, chills, etc) and a serological response to the *Cryptosporidium* antigen groups. Protective relationships were not found for diarrheal and gastrointestinal events or other illnesses and a serological response to the *Giardia* antigen. These results provide further evidence of the importance of protective immunity when attempting to understand the importance of endemic and epidemic waterborne illness. They also suggest that high levels of endemic *Cryptosporidium* infections commonly occur. Since these infections were observed in all three sites, it is unlikely that drinking water was a significant source of infection.

The possibility that low-dose exposure to pathogens offer protective immunity should be further evaluated and this evaluation should be given considerable priority. The question, "Will improving treatment of water that is already of high quality reduce or increase the occurrence of illness in the population served?" must be addressed.

REFERENCES

Birkhead, G., and R. L. Vogt. 1989. Epidemiological Surveillance for Endemic *Giardia lamblia* Infection in Vermont. *Am. J. Epidemiol,* 12: 762-68.

Blair, K. 1994. *Cryptosporidium* and Public Health. *Health and Environment Digest,* 8: 61-3.

Calderon, R.L. 2002. personal communication.

Colford J., T. Wade, S. Sandhu, C. Wright, S.F. Burns, A. Benker, S. Lee, A. Brookhart, M. VanDerLaan, D. Levy. 2003. A Randomized, Blinded, Crossover Trial of an In-home Drinking Water Intervention to Reduce Gastrointestinal Illness. (Abstract ISEE 691), *Epid,* 14(5):S137.

Chute, C.G., R.P. Smith, and J.A. Baron. 1987. Risk Factors for Endemic Giardiasis. *Am J Public Health,* 77: 585-87.

Craun, G.F. 1992. Waterborne Disease Outbreaks in the United States of America: Causes and Prevention. *World Health Statistics Quarterly* 45:192.

Craun, G.F., S.A. Hubbs, F. Frost, R.L. Calderon, and S.H. Via. 1996a. Waterborne Outbreaks of *Cryptosporidiosis*. *J. Am. Water Works Assn,* 90: 81-91.

Craun, G.F. 1996b. Waterborne disease in the United States. In, Craun GF, Ed. *Water Quality in Latin America: Balancing the Microbial and Chemical Risks in Drinking Water Disinfection.* ILSI Press, Washington, DC.

Craun, G.F. and R. Calderon. 1997a. Microbial Risks in Groundwater Systems: Epidemiology of Waterborne Outbreaks. *In Under the Microscope Examining Microbes in Groundwater*, Awwa Research Foundation, Denver, Colo.

Davis, L.J., J.L. Roberts, D.D. Juranek, S.R. Framm, and R. Soave. 1998. A Survey of Risk Factors for Cryptosporidiosis in New York City: Drinking Water and Other Exposures. *Epidemiol Infect,* 121:357-367.

Fraser, G.G. and K.R. Cooke. 1991. Endemic Giardiasis and Municipal Water Supply. *Am. J. Public Health,* 81:760-63.

Frost, F., G.F. Craun, and R.L. Calderon. 1996. Waterborne Disease Surveillance: What is it and Do We Need it? *J. Am. Water Works Assn,* 88: 66-75.

Frost, F.J., G.F. Craun, R. Calderon and S. Hubbs. 1997. *Cryptosporidium*: So Many Oocysts - So Few Outbreaks. December 1997 [opinion]. *J. Am. Water Works Assn,* 89:9-10.

Frost, F. 1998. Two-City *Cryptosporidium* Study. *American Water Works Association Research Foundation - Drinking Water Research* 8:2-5.

Frost F.J., A.A. de la Cruz, D.M. Moss, M. Curry and R.L. Calderon. 1998a. Comparisons of ELISA and Western Blot Assays for Detection of *Cryptosporidium* Antibody. *Epidemiol Infect,* 121(1):205-211.

Frost F.J., R.L. Calderon, T.M. Muller, M. Curry, J.S. Rodman, D.M. Moss, and A.A.de la Cruz. 1998b. A Two-Year Follow-up of Antibody to *Cryptosporidium* in Jackson County, Oregon Following an Outbreak of Waterborne Disease. *Epidemiol Infect,* 121(1):213-217.

Frost, F.J., T. Muller, G.F. Craun, D. Frasier, D. Thompson, R. Notenboom, and R.L. Calderon. 2000. Serological Analysis of a Cryptosporidiosis Epidemic. *Int J Epidemiol,* 29:376-379.

Frost, F.J., T. Kunde, T. Muller, G.F. Craun, and R.L Calderon. #1. A Comparison of Serological Responses to *Cryptosporidium* Antigens Among Populations Using Drinking Water Derived From Surface Versus Underground Sources. *Epidemiol. Infect, (in press).*

Frost, F. J., T. Muller, R. L. Calderon and G. F. Craun: #2. Analysis of Serological Responses to *Cryptosporidium* Antigen Among NHANES III Participants. *Ann Epi,* (in press).

Hellard, M.E., M.I. Sinclair, A.B. Forbes, C.K. Fairley. 2001. A Randomized Blinded Controlled Trial Investigating the Gastrointestinal Health Effects of Drinking Water Quality. *Environ Health Perspect,* 190: 773-778.

Hoxie, N.J., J.P. Davis and J.M. Vergeront. 1996. Cryptosporidiosis-Associated Mortality Following a Massive Waterborne Outbreak in Milwaukee, Wisconsin. Report of Wisconsin Department of Health and Social Services, Madison, WI.

Hunter, P.R., J.M. Colford, M.W. LeChevallier, S. Binder, P.S. Berger. 2001. Panel Summary From the 2000 Emerging Infectious Disease Conference in Atlanta, Georgia. *Emerg Infect Dis*, 7(3 Supple):544.

MacKenzie, W.R., N.J. Hoxie, M.E. Proctor, M.S. Gradus, K.A. Blair, D.E. Peterson, J.J. Kazmierczak, D.G. Addiss, K.R. Fox, J.B. Rose, et al.. 1994. A Massive Outbreak in Milwaukee of *Cryptosporidium* Infection Transmitted Through the Public Drinking Water Supply. *New Engl. J. Med,* .331:161-7.

Meinhardt, P.L., D.P. Casemore and K.B. Miller. 1996. Epidemiological Aspects of Human Cryptosporidiosis and the Role of Waterborne Transmission. *Epidemiol. Rev,* 18:118-136.

MLwiN statistical software, version 2.0. Centre for Multi-level Modeling, Institute of Education, London, England UK.

Morris R.D. and R. Levin. 1995. Estimating the Incidence of Waterborne Infectious Disease Related to Drinking Water in the United States. In: *Assessing and Managing Health Risks from Drinking Water Contamination: Approaches and Applications*, E. Reivhard and G. Zapponi, Eds.,IAHS Publication No. 233, pp. 75-88.

Moss D.M., C.L. Chappel, P.C. Okhuysen, H.L. Dupont, M.J. Arrowood, A.W. Hightower and P.J. Lammie. 1998. The Antibody Response to 27- 17- and 15-kDa *Cryptosporidium* Antigens Following Experimental Infection in Humans. *J. Infectious Dis,* 178:827-833.

Okhuysen, P.C., C.L. Chappel, C.R. Sterling, W. Jakubowski and H.L. DuPont. 1998a. Susceptibility and Serological Response of Healthy Adults to Reinfection with *Cryptosporidium parvum*. *Infection and Immunity,* 66:441-443.

Okhuysen, P.C., C.L. Chappel, C.R. Sterling, W. Jakubowski and H.L. DuPont. 1998b. Authors Reply. *Infection and Immunity,* 66(8):4008-9.

Payment, P., L. Richardson, J. Siemiatycki, R. Dewar, M. Edwardes and E. Franco. 1991a. A Randomized Trial to Evaluate the Risk of Gastrointestinal Disease Due to Consumption of Drinking Water Meeting Current Microbiological Standards. *Am. J. Public Health,* 81: 703-708.

Payment, P., E. Franco, L. Richardson, and J. Siemiatycki. 1991b. Gastrointestinal Health Effects Associated with the Consumption of Drinking Water Produced by Point of Use Domestic Reverse Osmosis Filtration Units. *Applied Environ. Microbiol,* 57:945.

Payment, P. J., Siemiatycki, L. Richardson, G. Renaud, E. Frano and M. Prevost. 1997. A Prospective Epidemiological Study of Gastrointestinal Health Effects Due to the Consumption of Drinking Water. *Int J Environ Health Res,* 7:5-31.

SAS, version 8.2. 2004 SAS Institute Inc., SAS Campus Drive, Cary, North Carolina 27513, USA.

Taylor, G.D., W.M. Wenman. 1987. Human Immune Response to *Giardia lamblia* Infection. *J. Infectious Dis,* 155(1):137-140.

ABBREVIATIONS

AWWA	American Water Works Association
AwwaRF	Awwa Research Foundation
CHRMC	Children's Hospital and Regional Medical Center
HIP	Health Improvement Partnerships
IRB	Institutional Review Board
LCF	Lovelace Clinic Foundation
USEPA	United States Environmental Protection Agency

IWA Publishing's authorised EU representative for General Product Safety Regulations is Diane D'Arras, 15 rue Duret, 75116 Paris, France, e-mail: safety@iwap.co.uk.

Printed and bound by CPI Group (UK) Ltd, Croydon, CR0 4YY
17/09/2026
02218798-0006